AUDEL®

MASONS AND BUILDERS LIBRARY VOLUME II

Bricklaying • Plastering • Rock Masonry • Clay Tile

by Louis M. Dezettel

revised by Tom Philbin

Macmillan Publishing Company
New York

Collier Macmillan Publishers
London

Macmillan Publishing Company
866 Third Avenue, New York, N.Y. 10022
Collier Macmillan Canada, Inc.

Library of Congress Cataloging-in-Publication Data

Dezettel, Louis M.
  Masons and builders library.
  Contents: v. 1. Concrete, block, tile, terrazzo—v. 2. Bricklaying,
plastering, rock masonry, clay tile.
  Includes index.
  1. Masonry. 2. Building. I.Philbin, Tom, 1934— II. Title.
TH5311.D48   1984   693'.1   83-22352
ISBN 0-672-23401-7 (set)
ISBN 0-672-23402-5 (v. 1)
ISBN 0-672-23403-3 (v. 2)

Macmillan books are available at special discounts for bulk purchases
for sales promotions, premiums, fund-raising, or educational use. For
details, contact:
        Special Sales Director
        Macmillan Publishing Company
        866 Third Avenue
        New York, N.Y. 10022

10 9 8 7 6 5 4 3 2

Printed in the United States of America

Foreword

There are many factors that must be taken into consideration when designing or constructing any type of building or small project around the house. Among them are terrain, geographical location, the type of material to be used, and the cost of the material.

Brick, concrete block, and structural clay tile have gained popular use as basic building materials for homes and industrial buildings. Many hours can be saved and labor costs cut when single-wall construction is erected with concrete block or structural clay tile. Masonary walls can be faced with brick, which not only beautifies the structure but adds insulation automatically.

So far, no automatic method for laying brick has been devised. It is a hand operation, depending entirely on the bricklayer's high degree of skill and efficiency. Years of experience have developed a pattern or technique in handling the trowel and knowhow in mixing the proper ingredients used in mortar.

The purpose of this book is to aid the builder, whether a person who does occasional masonry work around the house, a contractor, or an apprentice, by supplying the information needed to do a wide variety of masonry jobs properly. There is also information on plastering and installing plasterboard.

Louis M. Dezettel

Contents

CHAPTER 4

CHAPTER 5

CHAPTER 6

CHAPTER 7

CHAPTER 8

CHAPTER 9

CHAPTER 16

CHAPTER 1

Brick

As structural material, brick has reached a high state of art in strength, appearance, and other factors. Not only is it a popular basic material for building homes and industrial buildings, but it is highly regarded for its architectural and aesthetic possibilities (Fig. 1-1).

Methods for making basic brick have not changed much over the years. Even handmade adobe brick is still used to a limited extent. Yet brick is probably only second to wood in its long history of use. The art of brickmaking dates from very early times. Sun-dried, or adobe, brick were used thousands of years before the earliest recorded date of history, as given on a brick tablet of time of Sargon of Akkad, 3800 B.C., founder of the Chaldean Empire.

It was very natural for the dwellers along great rivers, such as the Euphrates and Tigris, to notice on the banks the sunbaked and irregularly cracked clay blocks which, after a little crude shaping, proved suitable for building a wall. Later, and about the

9

Fig. 1-1. For sheer beauty, few materials top brick, and fewer still beat its strength.

time when the Tower of Babel was built, the Chaldeans learned how to burn brick, thus converting the clay into a hard substance. In the time of Nebuchadnezzar (604–652 B.C.), the Babylonians and Assyrians had acquired the art, not only of making hard burnt brick, but also of beautifully enameling it. The Chinese claim great antiquity for their clay industries, but it is probable that the knowledge of brickmaking traveled eastward from Babylonia across all of Asia.

The Great Wall of China was partly constructed of brick, both burnt and unburnt. This was, however, built at a comparatively late period (210 B.C.), and there is nothing to show that the Chinese had any knowledge of burnt brick when the art flourished in Babylonia. The first brick buildings in America were erected on Manhattan Island in the year 1633 by a governor of the Dutch West Indies Company. The brick for these buildings was made in Holland, where the industry had long reached excellence. For many years brick was imported into America from Holland and also from England.

In America, burnt bricks were first made in New England in about 1650. The manufacture of brick slowly spread throughout the New England states. The Colonial days produced five types of brick architecture from New England to Virginia. In the nineteenth century, up to about 1880, American brick building was confined largely to the use of common brick for ordinary construction or for backing stone-faced walls. From that date to the present, a growing taste has demanded and secured artistic effect in the brick wall by the use of specially selected or manufactured brick of various shades and finish.

By definition, clay is a common earth of various colors, compact and brittle when dry, but plastic and tenacious when wet. It is a hydrous aluminum silicate generally mixed with powdered feldspar, quartz, sand, iron oxides, and various other minerals. The various kinds of clay are named for their suitability to a particular use—brick clay, fire clay, potter's clay, etc. All clays are the result of the denudation and decomposition of feldspathic and siliceous rocks and consist of fine insoluble particles that have been carried in suspension in water and deposited in geologic basins according to their specific gravity and degree of fineness.

These deposits have been formed in all geologic epochs from the Cambrian to the recent, and they vary in hardness from the soft and plastic alluvial clays to the hard, rocklike shales and slates of the older formations. The alluvial and drift clays which were used alone for brickmaking until modern times are found near the surface, are readily worked, and require little preparation, whereas the older sedimentary deposits are often difficult to work and require the use of heavy machinery. These older shales or rocky clays may be brought to a plastic condition by crushing and grinding in water, and they then resemble ordinary alluvial clays in every respect.

The clays or earth used in modern brickmaking may be divided into two classes according to chemical composition, as:

1. Clays or shales containing only a small percentage of carbonate of lime.

2. Clays containing a considerable percentage of carbonate of lime.

The first class consists chiefly of hydrated aluminum silicates, which is the true clay substance. Clays of this class usually burn to a buff, salmon, or red color. The second class, known as *marls*, may contain as much as 40% chalk. Marl burns to a sulfur yellow color which is quite distinctive. The color of brick depends on the composition of the material and the manner in which it is treated in the kiln. The chief colorant is the iron oxide in the clay, which does not show until the material has been heated, and which cannot be determined from an inspection of the raw material. It should be remembered that brickmakers often speak of clays as red clay, white clay, etc., according to the color of the brick made from them, without any reference to their color in the unburned state.

The strongest brick clays, or those possessing the greatest plasticity and tensile strength, are usually those which contain the highest percentage of the hydrated aluminum silicates. All clays contain, in greater or lesser amounts, some undecomposed feldspar. The most important ingredient other than the clay and sand substance is oxide of iron for color and, to a lesser extent, for hardness and durability. A clay containing from 5 to 8% oxide of iron will, under ordinary conditions of firing, produce a red brick. If the clay contains 3 to 4% alkalies or the brick is fired too hard, the color will be darker, approaching purple. Fenugenous clays generally become darker as they approach the fusion point. Alumina acts to make the color lighter.

All clays when heated sufficiently lose their plasticity and cannot regain it, so that on burning, they are converted into rigid bodies. Thus, when a clay is heated, the first effect is to drive off the water of formation. The clay then becomes dry but is not chemically changed; it does not cease to be plastic when cooled and moistened. On continuing to raise the temperature, the chemically combined water is separated and the clay undergoes a molecular change, which prevents it from taking up water again, except mechanically.

With the loss of the chemically combined water, the clay ceases to be plastic. On further heating, clays tend to undergo partial fusion. When this has occured to a sufficient extent for the fused material to fill the pores completely, the brick becomes impervious to water and is said to be *vitrified*.

The varieties of clays used in brickmaking are very numerous. Those which may be mentioned are:

1. **White burning clays.** These are composed chiefly of alumina, silver, and water, and are used to a very limited extent in brickmaking since cheaper materials are available.
2. **Marls.** Marls contain a considerable proportion of lime in the form of chalk or limestone. Bricks made from these materials are almost white; this is not due to the purity of the material, but to the combination of the iron oxide with the lime in the clay. Marls are easily fusible and give a characteristic effervescence when a little hydrochloric acid is poured on the surface.
3. **Loams.** Loams consist of clays containing a large proportion of sand, rendering them easier to work than tougher clays.
4. **Shales.** Shales are underrated clays which have been subjected to so much compression that they are almost semi-rock in characteristics. They have little plasticity. Red-burning materials obtained from impure shales are largely employed in brickmaking.
5. **Fire clays.** Fire clays are the refractory clays, or those capable of resisting very high temperatures in furnaces. They are, accordingly, used for making fire bricks, which are employed in lining boilers, furnaces, and fireplaces.
6. **Boulder clays.** Glacial action produces these clays, and they are distinguished from other clays by the number of rounded stones they contain. With careful selection and preparation, boulder clays make satisfactory common brick.

In addition to the foregoing classes, there are other designations of clays, such as:

1. **Brick earth.** The term brick earth is used to distinguish clays which can be made into brick without much mechanical treatment from the harder rock clays and shales which must first be ground. Accordingly, the machinery necessary to

manufacture brick from brick earth is reduced to a minimum.

2. **Fat clays.** Fat clays are those which are strong or plastic, containing a high percentage of true clay substance and a low percentage of sand. Such clays take up a considerable amount of water in tempering, dry slowly, shrink considerably, and lose their shape and develop cracks in drying and firing. Fat clays are improved by the addition of coarse, sharp sand, making the brick more rigid during the firing.

The presence of organic matter gives wet clay a greater plasticity. In some of the coal-measure shales the amount of organic matter is considerable, which renders the clay useless for brick making. Other impurities which frequently occur are the sulfates of lime, magnesia, chlorides, nitrates or soda, potash, and iron pyrites. All of these except the pyrites are soluble in water and are undesirable because they give rise to scum, which produces patchy color and pitted surfaces. The most common soluble impurity is calcium sulfate, which produces, in drying, a whitish scum on the brick surface. Such brick are of inferior quality because the scum becomes permanently fixed in burning.

Scumming is an important item in the manufacture of first-class brick. When a clay containing calcium sulfate must be used, a certain percentage of barium carbonate is usually added to the wet clay. This converts the calcium sulfate into calcium carbonate, which is insoluble in water, so that it remains distributed throughout the mass of the brick instead of being deposited on the surface.

Efflorescence is a white powder of crystallization caused by water-soluble salts, which are sometimes present in the brick or mortar. Water, such as rain, will sometimes leach the salts out of the brick, and they appear as white, powderlike patches on the surface. Currently, only a small percentage of brick produced in the United States contains enough salts to produce efflorescence.

HOW BRICK IS MADE

Because of the weight of the raw material and the finished product, transportation economy dictates that processing plants be located near the material source and near the markets. As a result, there are several hundred brick processors, both large and small, throughout the United States. Even a smaller plant requires a considerable capital investment in earth shovels, trucks, crushers, compressors, kilns, and drying sheds.

Clay is dug from nearby pits and trucked to the various plant locations. It is dumped into a crusher, at which point other chemicals may be added. The right amount of water is added to make the material into a heavy plastic "mud." The plastic material is then placed into a compressor. A compressor is shown in Fig. 1-2. The material, under great pressure, is squeezed out of a long rectangular-shaped tube, much as toothpaste is squeezed out of a tube. This can be seen near the center of the illustration. Near the left side of the illustration is a cutter, which cuts the extruded material into long bars, about 20 brick-lengths long. The cutter moves with the extruded material, making the operation a continuous one.

Fig. 1-2. Plastic clay material is placed in a compressor and squeezed into a long bar.

Each bar is moved over to a taut-wire brick cutter (Fig. 1-3). It cuts the bar into standard sized bricks. The brick is cut slightly oversize to allow for shrinkage during drying. The bricks are then piled, with air spaces between each brick, onto pallets with rail wheels. Because of the high compression of the extrusion, the

Fig. 1-3. A taut-wire cutting machine cuts the plastic bars into brick sizes.

plastic bricks are hard enough to support the weight of several layers. The pallets are moved by rail to one of many drying sheds (Fig. 1-4) where they are held for several days under moderate heat. This removes all moisture, leaving a brick that is dry but crumbles easily.

When dry, the bricks are moved to a kiln, where they are baked for 24 hours at high temperature. The kilns, or ovens, which are gas-fired (Fig. 1-5), are long and have doors at both ends. The pallets are wheeled into one end and removed from the other end. This makes the process almost continuous, with fresh brick going in one end as the cured bricks are removed from the other end. It is in the kiln that the bricks become hard.

Fig. 1-4. Drying sheds remove moisture from the bricks.

Fig. 1-5. Bricks are kiln-cured in long ovens for at least 24 hours.

17

Time and temperature determine hardness and water absorption. Common building brick is made of the natural clay or shale. Their outside color is not uniform. Although not normally used as face brick, some people prefer this effect for fireplaces and other purposes. Barium additives will prevent this discoloration and produce a single-colored brick for face brick. Other additives, or other types of clay or shale, produce colors other than the common "red" brick, as explained.

COLOR AND TEXTURE

The color of brick is the function of the material, the curing time, and heat of the kiln burning. Most color is caused by the amount of iron in the mix, which is part of the clay dug from the earth. The heat oxidizes the iron to form iron oxide, which is red.

The lighter colors (the "salmon" colors) are the result of underburning. If a higher and longer heat is applied, the brick will be harder; therefore, the lighter-colored, or underburned, brick will generally have less compressive strength and greater water absorption. These are used for decorative purposes as face brick, backed up by other tiers of harder brick. The sides are smooth as the plastic clay is extruded from the die. If the clay is not chemically treated, the brick cut from the extrusions forms the common brick, which is the backbone of all brick structures (Fig. 1-6). While color treatment is accomplished by chemical additives added to the clay and by the control of heat in the kiln, the texture is added after extrusion and before baking. Some deep-textured bricks are made by molding rather than by extrusion.

Brick textures may be stippled, water or sand struck, or have horizontal or vertical markings, or a host of other finishes. Fig. 1-7 shows two unusual finishes. Fig. 1-7A is vertically scored after the brick is cut from the extrusion. Fig. 1-7B looks like natural stone but is the result of molding brick in special forms.

STRENGTH AND WATER ABSORPTION

Strength is usually a measure of the compressive strength in pounds per square inch (psi) of the flat surface. Compressive

Fig. 1-6. Common brick has no chemical treatment for color. It is hard, has less water absorption, and is the basic structural brick.

strength and water-absorption qualities are a result of time and temperature in the kiln baking process, as mentioned before. Compressive strength is a serious consideration only when multistory walls of load-bearing types are to be built. Otherwise, the compressive strength of brick is usually more than adequate for normal use.

Most brick throughout the country is made with a compressive strength between 3000 and 7000 psi. About 8% of all brick is below and about 25% is above this figure. Water absorption is expressed in percent to total weight and is the percent of weight increase after the brick is soaked in water for a given period of time. Two soaking methods are used: either 24 hours in distilled water, or 5 hours in boiling water. Absorption is based on the following formula:

$$\text{Percentage of absorption} = \frac{100(B - A)}{A}$$

where A = weight of dry brick
       B = weight of saturated, or wet, brick

19

(A) Vertical scored.

(B) Molded brick in special form.

Fig. 1-7. Brick texture.

A dry brick is first weighed. It is then soaked in water for 24 hours or boiled in water for 5 hours. It is weighed again. Substituting in the above formula will give the percent of water absorption.

Another method of measuring water absorption is shown in Fig. 1-8. A glass cylinder is marked in arbitrary but evenly divided marks on the outside. In the example shown, water is poured into the cylinder to the 40 mark. When a brick is placed in the water, the water may rise to the 60 mark. This means 20 units of water have been displaced. After 24 hours the water may drop to the 55 mark. The five-division lowering of water means the brick has absorbed 25% of the displaced water.

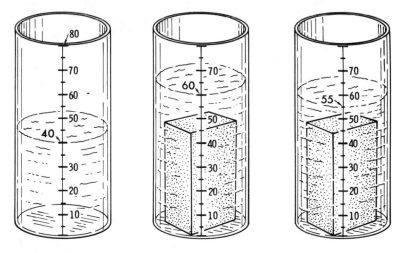

Fig. 1-8. One way of measuring water absorption in bricks.

Compressive strength and water absorption is considerably less important than good workmanship and proper mortar. For example, brick with a high water-absorption figure will soak some of the water out of mortar when laid up, but dampening the brick before using it will overcome this.

An important factor in providing years of good strength to brick structures is weathering. Freezing and thawing of water caught in the cracks and crevices of brick and openings left due to poor workmanship in the mortar can result in a faster deterioration than any other factor. In addition to good workmanship, brick texture selection and proper mortar pointing are important. Where weathering conditions are bad, deeply textured bricks should be avoided.

Fig. 1-9 is a weathering index for the United States. It is based on the number of times winter rains may go above or below the freezing temperature of 32°F.

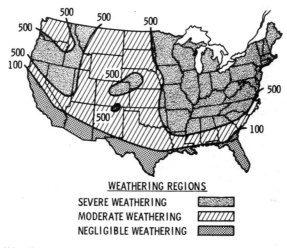

WEATHERING REGIONS
SEVERE WEATHERING
MODERATE WEATHERING
NEGLIGIBLE WEATHERING

Fig. 1-9. Weather map.

STANDARD BRICK SIZES

Nominal brick sizes are based on 4″ modules; actual sizes are slightly smaller to allow for mortar thickness. This is illustrated in Fig. 1-10. The 4″ squares are the module standard, and an "economy" brick is used to illustrate the concept. Generally the nominal size of "economy" brick is 4″ × 4″ × 8″, but its actual size is 3⅝″ × 3⅝″ × 7⅝″. This allows for a ⅜″ mortar thickness on all three sides of the brick.

The precise actual brick size varies from maker to maker, depending on the ability to control shrinkage during baking. Fairly close tolerances are maintained by most makers and differences are easily adjusted by mortar thickness at the time of

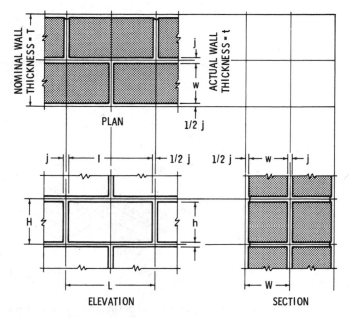

Fig. 1-10. The modular nominal dimensions of standard brick.

use. The most frequently used brick has a nominal size of 4″ ×
2⅔″ × 8″ in thickness, height, and length. A three-course high
layer consists of three bricks measuring two modules high, or 8″.
This size and its weight are handy for one-hand laying up.

Brick sizes are given names to identify their size and shape.
Fig. 1-11 shows the shape of those in popular use. Table 1-1 lists
the sizes of these bricks. These are nominal sizes, rather than
actual. Not all of these are available from all brick makers. Since
the user is dependent on the local suppliers, sizes should be
checked before planning any brick structures.

Locally available brick may or may not have hollow cores. The
hollow cores do not significantly reduce their strength or water-
absorption ability. The hollow cores are used primarily to reduce
weight and make life a little easier for the brick mason. Bonding
to the mortar is also improved because more surface is exposed
to the mortar. The nominal dimensions of hollow-core, load-
bearing, structural clay tile also follow the 4″ module design.
(These will be treated at some length in a later chapter.)

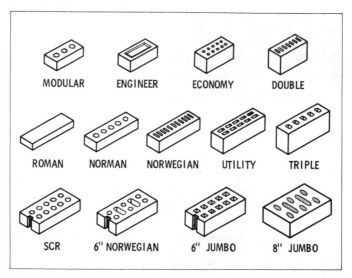

Fig. 1-11. Brick names indicate sizes as listed in Table 1-1. The brick may or may not have hollow cores.

Table 1-1. Nominal Brick Sizes

Unit Designation	Dimensions			
	Thickness	Height	Length	Modular Coursing
Standard Modular	4"	2²/₃"	8"	3C = 8"
Engineer	4"	3¹/₅"	8"	5C = 16"
Economy	4"	4"	8"	1C = 4"
Double	4"	5¹/₃"	8"	3C = 16"
Roman	4"	2"	12"	2C = 4"
Norman	4"	2²/₃"	12"	3C = 8"
Norwegian	4"	3¹/₅"	12"	5C = 16"
Utility*	4"	4"	12"	1C = 4"
Triple	4"	5¹/₃"	12"	3C = 16"
SCR brick†	6"	2²/₃"	12"	3C = 8"
6" Norwegian	6"	3¹/₅"	12"	5C = 16"
6" Jumbo	6"	4"	12"	1C = 4"
8" Jumbo	8"	4"	12"	1C = 4"

*Also called Norman Economy, General, and King Norman.
†Reg. U.S. Pat. Off., SCPI.

ADOBE BRICK

The oldest form of brick is the adobe brick, dating back to 3800 B.C., as mentioned at the beginning of this chapter. It was also the principal method of dwelling construction by the Indians of the Southwest and by the first Mexicans settling into the southwest territories, then owned by Spain. It is still in limited use.

Adobe is made of mud, finely chopped straw, and water. Clay occurs in good percentage in the natural earth of many regions in the Southwest and forms the bond which holds the materials into a solid brick when dry. There are no large-scale processors of adobe brick, which makes the manufacturing of adobe brick a do-it-yourself project. The well-mixed ingredients are placed in molds and are sun-dried until hard.

It has been assumed that an adobe home is cool in the summer and warm in the winter. Tests have shown that it takes a 3-ft.-thick adobe wall to equal the insulation qualities of a frame house with $2'' \times 4''$ construction. Adobe absorbs water easily, unless specially treated, and is a hazard in areas subject to flooding.

Because of their weight (30 lb. or more per brick) and limited demand, it is not economical to ship adobe brick more than limited distances or to use professional masons for adobe home construction. While there are a few home builders who would be willing to construct adobe homes, the home building that remains is confined to areas where the homeowner's "sweat-equity" is in the work he puts into it by making the brick, putting in his own footings, and laying up the brick.

CHAPTER 2

Mortar

The weakest part of a brick wall is the mortar used to bond the bricks together. Well-baked brick has tremendous compressive strength and low water-absorption qualities. Not so with mortar—its quality depends on the proportions of the ingredients and the workmanship in using it.

INGREDIENT PROPORTIONS

Mortar is basically concrete and could have all of the durability qualities of concrete as described in the chapter on concrete in Volume I. Fundamentally, the ingredients would be the same, except for one factor: workability. Because mortar is applied to brick and the brick is laid up into a wall by hand, the operation is much slower than pouring concrete. Mortar must be treated to reduce its initial setting time. Because it is troweled to provide a given thickness of mortar joint and bricks must hold many layers or courses above it without settling down by weight, it must hold a shape. These requirements make mortar different from concrete.

Mortar workability is a function of the amount of hydrated lime used in the mix. Final strength is greatest when there is no hydrated lime. Table 2-1 shows the relative strength of mortar, varying with the amount of hydrated-lime content and the flow, which is a measure of the water content. The flow after suction is the flow after 1 minute of contact with the brick—the brick withdrawing some of the water from the mortar. Note the greater strength where the difference between the initial flow and the flow after suction is the least. This points up the importance of using brick with a minimum of water absorption. It is very important to first wet the brick if there is some suspicion the brick has high water-absorption qualities.

Masonry cement is portland cement with the necessary additives to improve workability. Masonry cement does not require the addition of hydrated lime. For increased strength, masonry cement may be added to straight portland cement. Table 2-2 shows the recommended mixes for two general services.

Mortar is also known by a type number for specific service needs. The following are the type numbers, the proportions of ingredients recommended, and the service for which they are intended:

Table 2-1. Compressive and Tensile Strength of Mortars

Mortar No. and Mix†	Initial Flow, Percent	Flow After Suction, Percent of Initial Flow	Strength, psi*	
			Tension	Compression
1	100	87	457	5492
1:¼:3	120	87	425	5153
	133	87	420	4830
2	100	89	300	2758
1:½:4½	120	88	277	2408
	133	88	268	2175
3	100	92	180	1173
1:1:6	120	93	165	905
	133	91	145	793

*Proportions:Cement, lime, sand by volume.
†Tension specimens, briquets; compression, 2-in. cubes; both cured in water, tested at 28 days.

Type M—1 part portland cement, ¼ part hydrated lime or lime putty, 3 parts sand; or 1 part portland cement, 1 part Type II masonry cement, and 6 parts sand. This mortar is suitable for general use and is recommended specifically for masonry below grade and in contact with the earth, such as foundations retaining walls, and walks.

Type S—1 part portland cement, ½ part hydrated lime or lime putty, 4½ parts sand; or ½ part portland cement, 1 part Type II masonry cement and 4½ parts sand. This mortar is also suitable for general use and is recommended where high resistance to lateral forces is required.

Type N—1 part portland cement, 1 part hydrated lime or lime putty, 6 parts sand; or 1 part Type II masonry cement and 3 parts sand. This mortar is suitable for general use in exposed masonry above grade and is recommended specifically for exterior walls subjected to severe exposure as, for example, on the Atlantic seaboard.

Type O—1 part portland cement, 2 parts hydrated lime or lime putty, and 9 parts sand; or 1 part Type I or Type II masonry cement and 3 parts sand. This mortar is recommended for load-bearing walls of solid units where the compressive stresses do not exceed 100 psi and the masonry will not be subjected to freezing and thawing in the presence of excessive moisture.

Table 2-2. Recommended Mortar Mixes
Proportions by Volume

Type of Service	Cement	Hydrated Lime	Mortar Sand in Damp, Loose Condition
For ordinary service	1—masonry cement* or 1—portland cement	— 1 to 1¼	2 to 3 4 to 6
Subject to extremely heavy loads, violent winds, earthquakes, or severe frost actions. Isolated piers.	1—masonry cement* plus 1—portland cement or 1—portland cement	— 0 to ¼	4 to 6 2 to 3

*ASTM Specifications C91

The information in Table 2-3 is based on proportions to 1 cu. ft. of cement, whether regular portland or masonry cement is used. Cement is purchased by the sack, which contains 1 cu. ft.

SAND

The aggregate (sand) used in mortar should be well graded and clean. It must not contain any organic material, such as salts or alkalies. These will weaken the mortar. All sands must pass the ⅛" square sieve and be gradually graded from that size down, but not smaller than that which passes the No. 100 sieve. The better the grading of sizes between those two, the better the voids between sand grains will be filled and the less the amount of cement that will be needed. It is a matter of economy rather than strength.

WATER AND PLASTICITY

The less the lime content of mortar, the greater the strength; but lime improves workability. The less the amount of water used (down to a limit), the greater the strength of the mortar will be, as it is with concrete. But water, too, is important to workabil-

Table 2-3. Quantities of Materials per Cubic Foot of Mortar

Mortar Mixes			Quantities			
Cement Sack*	Hydrated Lime, cu. ft.	Sand† cu. ft.	Masonry Cement Sack	Portland Cement Sack	Hydrated Lime, cu. ft.	Sand, † cu. ft.
1 Masonry Cement	—	3	0.33	—	—	0.99
1 Portland Cement	1	6	—	0.16	0.16	0.97
1 Masonry cement plus 1 Portland Cement	—	6	0.16	0.16	—	0.97
1 Portland Cement	¼	3	—	0.29	0.07	0.86

*1 sack masonry cement or portland cement = 1 cu. ft.
†Sand in damp, loose condition.

ity. For this reason, there is no recommended ratio of water to cement, as there is for concrete.

The amount of water in mortar is more a matter of experience than of any rule. Since some water is lost due to absorption by the brick, some allowance is made for this. More important is the effect on the ease with which mortar is applied to brick when a wall or other structure is laid up. The water content must not be so great that the mortar slides off the trowel when picked up or oozes off the end of a brick when it is laid up.

If too much water is absorbed by the brick, it will leave too little for proper hydration of the cement, and the mortar will lack strength. A rough test for excessive water absorption in brick is as follows: Sprinkle a few drops of water on the flat side of a brick. If the water is absorbed in less than 1 minute, absorption is too fast. If this is the case, a hose should be used to water the pile of bricks to the point where water runs off on all sides. However, do not use the brick immediately. It must be dry on the surface before use. Any drinking water is usually good for use with mortar. Avoid water with alkalies or salt.

HOW TO MIX

From a time- and labor-saving standpoint, mixing is best done in a power mixer of the kind used to mix concrete. A revolving drum powered by an electric or gasoline engine is used.

Small quantities may be mixed by hand. A flat board structure or box is used. The board or box must be tight to prevent the loss of water. If built well, it may be used over and over again. A shovel for turning the dry ingredients and a hoe for mixing when water is added are essential. For either method, mix the dry ingredients first. Turn the ingredients over three or four times until thoroughly mixed. Add water to the dry ingredients until the proper plasticity is obtained. Use a hose or bucket. If mixing is done by hand, make a depression in the center of the dry ingredients. Pour water into the depression. With a hoe, bring the dry ingredients at the edge over into the depression and stir. If more water is needed, make another depression and add more water, as before.

All the ingredients, including the water, must be thoroughly mixed to a mudlike consistency. Usually there is enough water when the hoe can be shaken free of mortar mix, but it must not be too wet, as explained before.

RETEMPERING

Retempering means adding water to mortar when it becomes too stiff while in use. Water may be added to keep it plastic if stiffening is due to water evaporation. Mortar that has stiffened because it has begun to set must not be retempered but must be thrown away and new mortar mixed.

A rule of thumb to decide if stiffening is due to evaporation or hydration is as follows: Mortar beginning to stiffen before 2½ hours at temperatures of 80°F or higher may be retempered. At temperatures below 80°F, the time may be extended to 3½ hours. Any mortar older than these allowed times should be discarded since it will have begun to set due to hydration, and retempering will only weaken the mix.

COLORING MORTAR

Masonry cement may be purchased in a number of colors. Since colors are premixed under careful control, the color from batch to batch will be consistent. When mixing your own, only mineral colors must be used with white portland cement. The following are the color ranges and how they are obtained:

Pink to red—Red iron oxide
Browns—Brown iron oxide
Yellow to buff—Iron hydroxide
Gray to blue slate—Manganese dioxide, black iron oxide, or 1% to 2% carbon black
White—White cement, white sand, and white stone

Mix all dry ingredients first. Mixing must be thorough. When the water is added, stir until there are no streaks visible.

EPOXY

No mortar is as strong as the bricks which are used. However, where strength is needed, as for load-bearing walls, an epoxy additive may be used. Epoxies and the amounts to be used vary with the manufacturer. Your supplier can tell you what to buy and how to use it. Epoxy adds somewhat to the cost of mortar and should be used only where the extra strength is needed. It can make mortar even stronger than the brick used and has tremendous bonding power to the brick.

CHAPTER 3

Tools

The number of tools needed by a brick mason are not numerous or expensive. The essential ones are shown in Fig. 3-1. In addition to these, some masons carry a hoe and shovel for preparing and handling the mortar, a large carpenter's right angle, other sizes of spirit levels, and jointing tools.

Many contractors will find it a worthwhile investment to purchase a power-driven mixer for the mortar. Contractors for large projects may also invest in a forklift, or elevator, for carrying brick to a higher level. The old hod-carrying days, when men used to carry brick and mortar on their shoulders up a ladder are practically gone.

SPIRIT LEVEL

The most important aspect of laying brick is the first, or starter, row. To lay a perfectly vertical wall, with true corners and equal

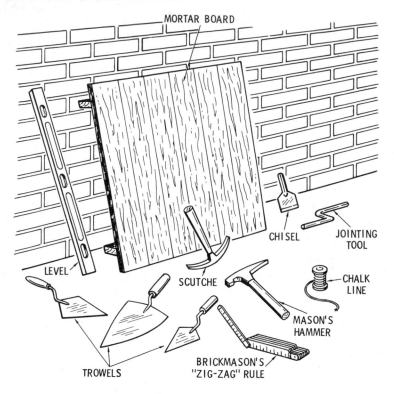

MORTAR BOARD

CHISEL

JOINTING TOOL

LEVEL

SCUTCHE

CHALK LINE

MASON'S HAMMER

TROWELS

BRICKMASON'S "ZIG-ZAG" RULE

Fig. 3-1. Principal tools used by the brick mason.

spacing of mortar joints, laying of the corners and first course are most important.

A brick wall starts with two corners raised part way. The distance between corners must be precisely measured to follow the 4″ module concept. The first course of brick is laid on a bed of mortar over a concrete footing. The spirit level is used to make sure this first course is perfectly horizontal, as it will affect all the courses above it. Fig. 3-2 illustrates a magnesium spirit level, with glass bubbles for both horizontal and vertical checking. Included is a 45° bubble for those few occasions where it may be needed. Longer levels may be made of wood, aluminum, or magnesium.

Several courses of each corner are laid, and the spirit level is used to make sure they are perfectly vertical. A large carpenter's

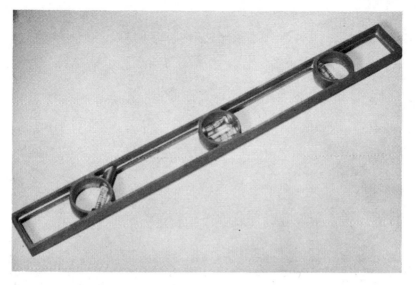

Fig. 3-2. A medium-size, magnesium spirit bubble level.

square is used to make sure adjacent sides are square. From that point on, a taut line is used between corners to set the mark for succeeding courses of horizontal brick rows. An occasional check with the spirit level ensures that rows are horizontal.

MEASURING

Essential to a good start is accurate measurement, as explained before. Fig. 3-3 shows two popular types of measuring devices. One is a folding metal rule, opening to 72″; the other is a 50-ft. tape with a hand rewind.

The tape is used to establish the corners based on multiples of the 4″ module. The rule sets the position of the taut line for each course as the wall is laid up. The vertical measurements also follow the module concept, using the standard brick sizes described in Chapter I of this volume. For example, three courses of the standard module brick will lay up a wall 8″ high, including the thickness of the mortar.

37

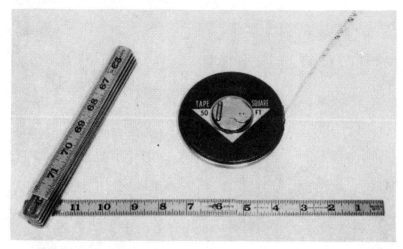

Fig. 3-3. Two popular measuring tools: the metal folding rule and a 50-ft. rewind measuring tape.

T-ANGLE

The adjustable T-angle, shown in Fig. 3-4, is used to accurately transfer one angle to another. For example, a corner other than 90° may be transferred to another corner to match. The tool has a thumbnut for adjusting the angle to the first corner. Tightening the screw holds the angle for transfer to another corner, either as an identical angle or as one that is 180° greater. Thus, a third adjacent wall will come out parallel to the first.

TAUT LINE

The next and most used measuring tool is the taut line. It is fastened from corner to corner to establish the height of the next course of brick. The best line is made of linen, approximately 84 ft. long (Fig. 3-5). As shown in Fig. 3-6, the line is drawn tight from corner to corner, mortar is laid along the lower course of brick, and a new row of brick is pushed or tapped into place to

Fig. 3-4. An adjustable T-angle.

Fig. 3-5. A good grade of linen taut line is important in accurately laying a brick wall.

match the line. Fig. 3-7 shows masons laying in a mortar line just below the taut line. The next step will be placement of brick for the next course.

TROWELS

The large trowel (Fig. 3-8, top) is the most used tool in the mason's kit. It is in constant use in handling mortar when laying up a wall. The next smaller size is convenient for buttering the ends of brick when the mason has a helper who does nothing but butter the brick. The smallest size is ideal for tuck pointing, which is repairing mortar in an old brick wall. Fig. 3-9 shows a medium-size trowel, ideal for the do-it-yourself home handyman with only occasional brickwork to do.

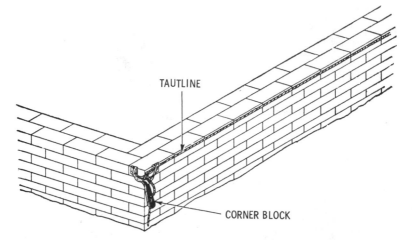

TAUTLINE

CORNER BLOCK

Fig. 3-6. The use of a taut line from corner to corner as a guide for the next row of brick.

Fig. 3-7. Masons laying a mortar line on which a row of brick will be placed even with the taut line.

The large trowel should be the best that money can buy. Being in constant use with abrasive material, it must stand up under long periods of use. It is also used by experienced masons to cut brick into smaller sizes by hitting the brick with the edge of the

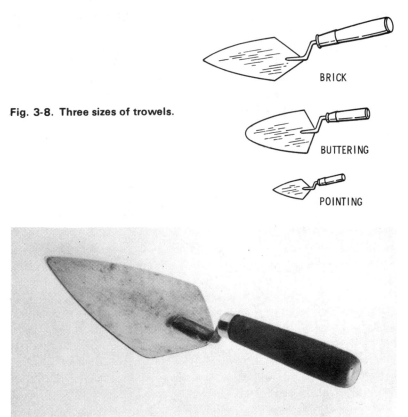

Fig. 3-8. Three sizes of trowels.

BRICK

BUTTERING

POINTING

Fig. 3-9. A small trowel—good for "do-it-yourselfers."

trowel. If it is not made of well-tempered steel, the edges will soon become badly marred. Handles on the trowel are usually wood. The end is used to tap the brick into position, but the tapping is light and the wood is not badly damaged even with long use.

BRICK CUTTERS

The module method of construction reduces the need for placing a short or half brick. Sometimes circumstances require a brick less than the standard size, and this is done by cutting. A sharp blow with the edge of a trowel, in the hands of an expert, will break a brick pretty close to the size desired. Hard brick is not easily cut this way. Two tools are used for cutting brick.

The mason's hammer (Fig. 3-10) has a curved, chisel like edge on one side. With practice, one soon learns to cut brick close to the desired size with a few whacks of the chisel side. This side is

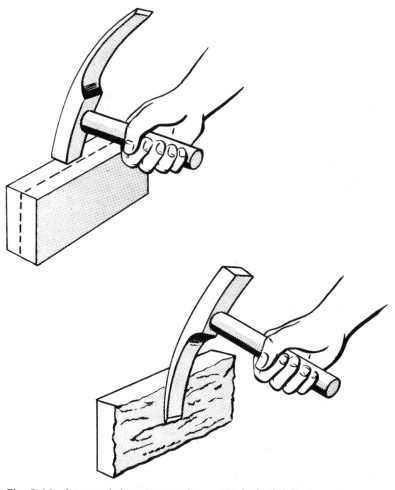

Fig. 3-10. A mason's hammer used to cut and trim brick.

also used to clean the face or edge of the brick if the break is rough. The chisel side of a mason's hammer is used to knock off old mortar when an old wall is being repaired with new sections.

The other side of the hammer has a square flat side somewhat like a hammer. It is *not* used as a hammer in conjunction with a chisel. A wide-edged chisel (Fig. 3-11) is the most accurate method for cutting brick, but it should be used with a wood mallet or a hand-sledge-type hammer, not the mason's hammer. Often the chisel is used to score a line at the breakpoint, and the brick is subsequently broken into two parts with the mason's hammer.

Large contractors use a carbide-toothed power saw to cut brick to the exact size, with square sides and with greater overall economy than can be done by hand. A brick broken into a piece smaller than its standard size is called a *bat*. A half brick is called a *half bat*. Anything between a half size and full size, but not including full size, is called a *three-quarter bat*.

THE MORTAR BOARD

Whether mortar is mixed by hand or by a power mixer, quantities of it are carried and placed on a mortar board for use by the mason when laying up a brick wall. Mortar boards are generally wood and easily homemade.

Sections of 1″ × 8″ wood are cleated together into a board about 28″ × 30″ (Fig. 3-12). The cross cleats are of the same material. Close-grained hardwood should be used. It is important that the wood absorb only a minimum amount of water from the mortar, and that there is no leak through the edges of the boards. Plane the edges true or use wood that is not warped. The mortar boards are placed on platforms for easy access to the mason or are constructed with legs. The less the mason needs to bend over, the less tiring the job and the more efficiently he can work.

In confined spaces, the mortar may be contained in a box smaller than the regular mortar board. Sides make it possible to hold a fair quantity of mortar. However, a flat board with no

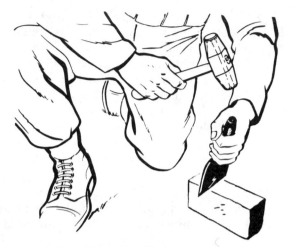

Fig. 3-11. Scoring brick with a wide chisel.

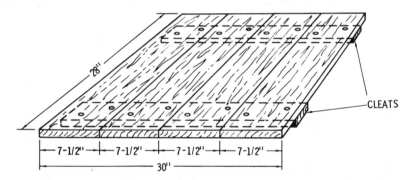

Fig. 3-12. Typical homemade mortar board.

sides provides easier access for the mason, and there is no waste from the mortar caught in corners.

JOINTING TOOL

Any curve-faced tool of metal may be used to press the mortar into the joints of brick after the excess has been struck off. Tools

Fig. 3-13. A jointing tool used to press the mortar into a tight concave joint.

are made in various sizes for different joint thicknesses. They are S-shaped and curved on one side. A bent, solid, round, metal rod may also be used but is adds unnecessary weight. Fig. 3-13 shows a jointing tool used to put a concave joint into brick mortar.

CHAPTER 4

Bonding

To bond means to bind or hold together. Bonding is important in brick construction to make a solid and secure structure. There may be three different meanings to the word "bond," as it refers to masonry. These are:

Structural bond. The interlocking of masonry units by overlapping bricks or by metal ties.
Pattern bond. Interlocking and overlapping brickwork following a fixed sequence. Pattern bonds for structural purposes have become standardized and are given names. Some pattern bonds are used for appearance purposes only or combined to provide a special appearance plus structural bonding qualities.
Mortar bond. The adhesion of the mortar to the masonry or to steel reinforcement ties placed in the masonry. Mortar alone is not strong enough to provide sufficient bonding for secure structures.

In overlapping brick construction, most building codes require that no less than 4% of the wall surfaces consist of headers, with the distance between headers no less than 24″ vertically or horizontally. Headers are bricks laid with their longest dimension perpendicular to the front, or facing, tier of bricks that overlaps into the second row, or tier, behind, for a double-thick wall. Steel ties are in wide use, with codes requiring at least one tie for each 4 ½ sq. ft. of surface.

TERMS APPLIED TO BONDING BRICK

Common names have been applied to bricks, depending on their position in the structure. Fig. 4-1 is a sketch of their names and positions.

Most wall construction is done with *stretchers*. Stretchers are bricks with their largest surface horizontal and longest edge facing out. *Headers* have their smallest dimension facing out and are

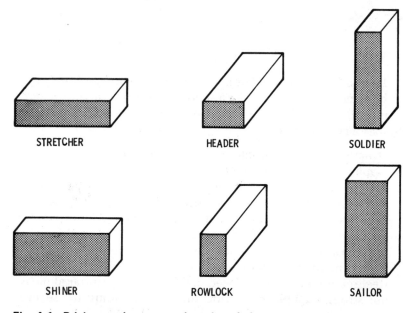

STRETCHER HEADER SOLDIER

SHINER ROWLOCK SAILOR

Fig. 4-1. Bricks are given names based on their use and position.

48

used to interlock with a wall of brick behind. A *soldier* is a brick standing on its end and is often used as the top course of a wall, forming a sill for the support of ceiling or roof joists (Fig. 4-2). When a soldier is placed with its largest face forward, it is often called a *sailor*. A *rowlock* is a header standing on edge. A rowlock with its largest surface facing out is sometimes called a *shiner*. Rowlock placement reduces the number of bricks in a double-thick wall, but it is not as strong.

Fig. 4-2. How soldiers are used for the top course as a sill. Generally their main purpose is decorative

Fig. 4-3 illustrates the method used to bond the bricks shown in Fig. 4-1. Each layer of brick is called a *course*. Stretchers are laid overlapping and provide the interlock between courses. Rowlocks are placed as stretchers or headers for interlocking. The dimensions are such as to leave a cavity between vertical tiers. *Wythe* (or withe) is the term generally applied to the grout between brick faces. The grout used is a thinner mortar to permit its flowing completely into every crevice and irregularity of the brick surfaces. Wythe also refers to the wall between flue cavities of a chimney.

49

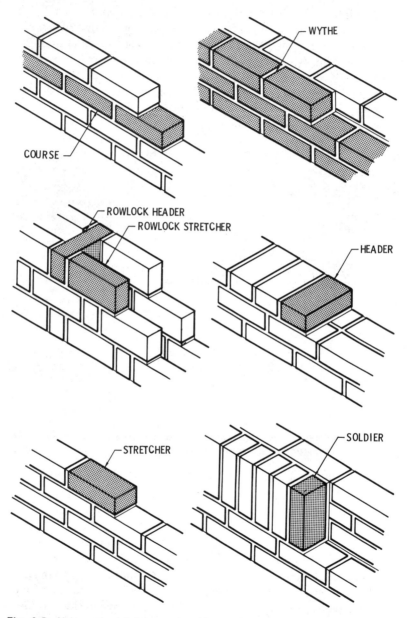

Fig. 4-3. How various bricks are used in overlap bonding.

OVERLAPPING AND THE MODULE CONCEPT

As mentioned in a previous chapter, nominal brick dimensions are based on the 4″ module concept. The face and long edge are a multiple of 4″, while the thickness is a simple fractional part of 4″ (usually ⅓ of 8″). Actual dimensions are slightly less to allow for mortar thickness. Modular dimensions permit placing brick in walls with both stretchers and headers to result in a fixed pattern in appearance, without the need for cutting brick into small pieces, except in a few cases.

Fig. 4-4 shows how a stretcher is twice the width of a header and, conversely, a header is one-half the width of a stretcher.

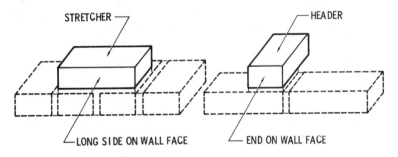

STRETCHER — ⌐   ⌐— HEADER

⌐— LONG SIDE ON WALL FACE   ⌐— END ON WALL FACE

Fig. 4-4. The nominal length of a brick is just twice that of its width.

MODULAR LAPPING

Modular dimensions have resulted in standard overlapping techniques, and the expressions ½ lap, ¼ lap, and ⅓ lap have developed.

The most common method of laying brick with stretchers, most used in walls, is the ½-lap method. This is clearly shown in Fig. 4-5. One-half the brick length in each successive course overlaps ½ the brick length below it. On occasion, for a change of pace in appearance, ⅓ lap is used. One-third of the brick length overlaps ⅔ of the brick below it. However, with each reduction in overlap, the strength of the mortar bonding is reduced. The limit of reduction is no overlap at all, sometimes used for garden

51

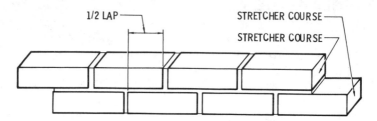

Fig. 4-5. Showing ½-lap bonding commonly used in stretcher courses.

walls. To retain strength in such cases, a number of metal ties are used for bonding.

The use of ¼ lap is generally with headers, or headers and stretchers. Fig. 4-6 shows two courses of headers with ¼ lap. This provides the maximum bonding for the shorter width of headers. The bonding is between half-widths of headers, but the ¼ lap is the width of a stretcher. Fig. 4-7 shows the ¼ lap used between courses of headers and stretchers. This is the method commonly used for a full course of headers. Where headers alternate with stretchers in the same course, the ¼ lap will look like Fig. 4-8. Note that headers have a full ½-lap bonding surface with facing stretchers above and below, but stretchers are reduced to ¼- and ¾-lap bonding. In rowlock construction, with alternate headers and stretchers, the result is ⅓ lap at all bonding surfaces (Fig. 4-9).

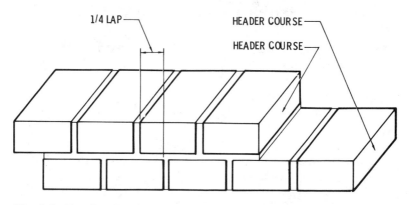

Fig. 4-6. Best bonding for courses of all headers is the ¼ lap.

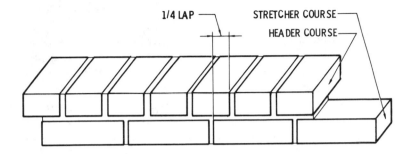

Fig. 4-7. A header course with ¼ lap over a stretcher course.

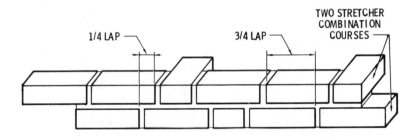

Fig. 4-8. Headers and stretchers in the same course result in a ¼ lap and a ¾ lap.

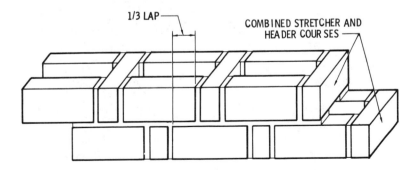

Fig. 4-9. In rowlock construction the result is ⅓-lap bonding.

53

STRUCTURAL PATTERN BONDS

When brick is laid with good bonding in mind, a pattern is formed, which, when repeated consistently, results in an appearance that is pleasing as well as strong. This is especially true when header bricks are involved in the structure.

Over the years standard patterns have been established, and each is given a name for identification. About six patterns are in wide general use. They are illustrated in Fig. 4-10.

RUNNING BOND

1/3 RUNNING BOND

6TH COURSE HEADERS
COMMON BOND

6TH COURSE FLEMISH HEADERS
COMMON BOND

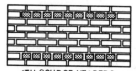

DUTCH CORNER · ENGLISH CORNER
FLEMISH BOND

ENGLISH CORNER   DUTCH CORNER
ENGLISH BOND

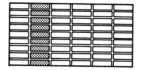

STACK BOND

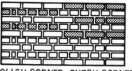

ENGLISH CORNER   DUTCH CORNER
ENGLISH CROSS OR DUTCH BOND

Fig. 4-10. The most popular bonding patterns in general use.

Running Bond. All bricks are laid in ½-lap stretcher bonding, without headers. A variation is the ⅓ running bond, as mentioned before (Fig. 4-11). Because there are no headers for bonding to a second thickness of brick behind, metal ties must be used between thicknesses. The metal ties allow for an air cavity between sections. The air space provides extra insulation against the transmission of heat through the brick wall. If strength is more important than insulation, the cavity may be filled with concrete. The running bond is also used for homes of brick-veneer construction, which are homes of 2″ × 4″ framing and a single layer of facing brick, essentially for appearance. A common method of home construction in some areas is an all-frame home with brick veneer in front and wood or stucco on the other three sides.

Common, or American, Bond. Perhaps the most frequently used bond pattern is the common, or American, bond. The pattern consists of several courses of stretchers only and headers every fifth, sixth, or seventh course, depending on the needs

Fig. 4-11. Here ⅓-lap running bond is used in a single-tier, brick-veneer form.

for structural strength. A continuous course of headers is used. A variation is the Flemish header course in which headers alternate with stretchers.

In the common-bond pattern, each row of headers must start their corners with a ¾ length of brick, to come out even with a symmetrical pattern, on the 4″ module system. The half-size brick appearing at the ends of some stretcher courses is a full stretcher starting the adjacent wall around the corner.

Flemish Bond. This bond pattern is obtained by using alternate header and stretcher bricks with each header on one course centered with a stretcher in the course below and above. Where the strength of so many headers is not needed, many of the headers shown can be clipped brick or ½-bat size. However, there is no reduction in total brick used, and labor can be saved by not clipping the brick.

English Bond. This pattern is formed with alternate courses of running stretchers and running headers. As with the Flemish bond, many of the header bricks may be clipped if preferred.

English Cross, or Dutch, Bond. Similar to English bond, except the stretchers are spaced so each header faces the middle of a stretcher on one side and a joint between stretchers on the other. The joints form a series of overlapping X's, thus the name English "cross."

Stack Bond. Bricks are stacked vertically and horizontally with no overlap to form a stacked bond. Bonding strength between bricks and to the layer of brick behind is by means of metal straps or ties. This pattern is seldom used for a load-bearing wall, where the extra strength of overlapping is a must. Where stack bond is specified for load-bearing walls, a liberal use of steel reinforcing bars is important. This is also called *block pattern.*

CORNERS

Fig. 4-10 shows two types of corners for the English, Flemish, and Dutch bond patterns. In the Dutch corner, the corners start with a ¾ bat. In the English corners, they start with a ¼ bat. The ¼ bat, however, must never be placed at the corner, but at least

4″ from the corner. These corners allow for proper spacing to make the pattern come out as intended.

QUOINS

Bricks that are cut for use on corners are called *quoins*, a word that is slowly passing out of popular use. Fig. 4-12 illustrates the various quoin shapes and how they are used. The diagonally clipped, or king, quoin will have the appearance of a ¼ bat

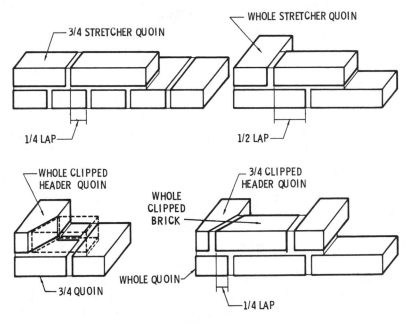

Fig. 4-12. Clipped bricks used at corners are generally called *quoins*.

visible in the brick face but the structural strength of a larger size. Fig. 4-13 shows two methods of making the English corner. With a king closure or quoin, the brick will have the appearance of a ¼-lap brick at the very corner without the attendant weakness.

Fig. 4-14 shows the relation of the cut brick to the whole brick. Brick is easily cut by first scoring it with a chisel having one

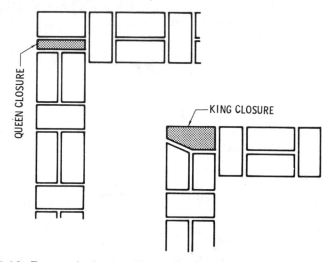

Fig. 4-13. Two methods of making an English corner.

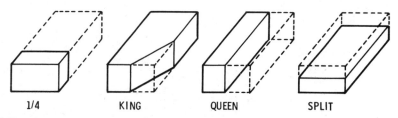

Fig. 4-14. Cut brick used for closures and corners.

straight side and one beveled side (Fig. 4-15). After scoring, a whack with the flat of the mason's hammer head will break the brick off even with the scored lines. It takes experience to obtain the precision needed for making clean breaks.

CLOSURES

All brickwork starts with the corners. The last brick placed to complete a course is called a *closure*, or *closer*. Most often the closure is a full stretcher somewhere in the middle of the course.

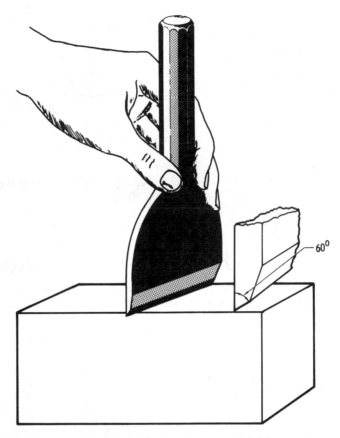

60^0

Fig. 4-15. The type of beveled chisel used to score brick for cutting.

For many structural pattern bonds, closures require cutting brick as described above. Fig. 4-16 shows some of the frequently used closures, in addition to the full stretcher.

GARDEN WALL

A variation of the Flemish pattern bond was given the name *garden-wall pattern.* If two or more stretchers are used between headers, the back and front of the brick wall will have the identi-

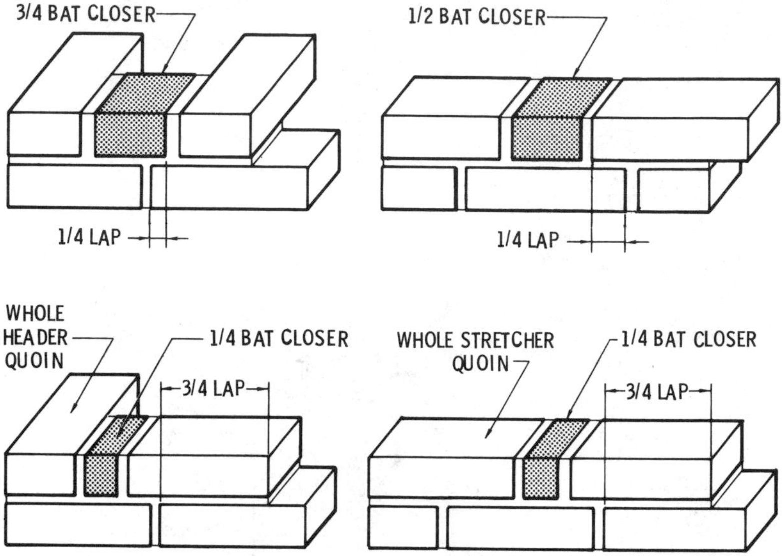

3/4 BAT CLOSER

1/2 BAT CLOSER

1/4 LAP

1/4 LAP

WHOLE
HEADER
QUOIN

1/4 BAT CLOSER

3/4 LAP

WHOLE STRETCHER
QUOIN

1/4 BAT CLOSER

3/4 LAP

Fig. 4-16. Variations of cut brick used as closures.

cal appearance. An early use of this design was for garden walls. Fig. 4-17 shows how brick is laid for two-, three-, and four-stretcher, garden-wall construction.

Figs. 4-18 and 4-19 show two garden-wall patterns with intentional appearance design. Note the free use of $1/4$-bat closures to make the diagonal designs come out even. Part of the design results from the use of bricks with differing textures and shades of colors. See Chapter 8 for more information on special design patterns.

METAL TIES

There are certain conditions in which the use of header brick for bonding to the rear brick wall section is not appropriate. Metal ties are then used instead of header brick.

Fig. 4-20 illustrates the method of bonding a brick-facing wall to a tile backing. Also shown are some of the shapes of metal ties available. While clay-tile sizes follow the 4″ module design con-

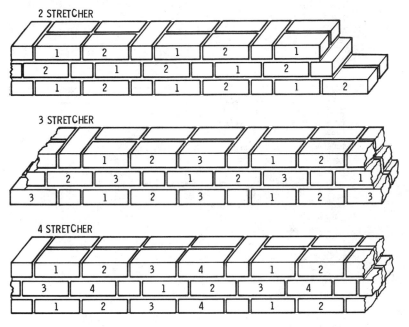

2 STRETCHER

3 STRETCHER

4 STRETCHER

Fig. 4-17. A variation of the Flemish pattern which is called *garden wall*. It has the same pattern on each side.

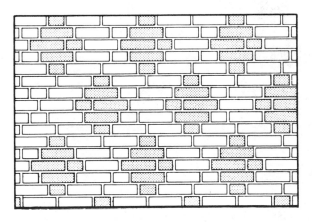

Fig. 4-18. Garden-wall design emphasizing diagonal lines.

61

cept, their larger size eliminates the use of headers for bonding. Metal ties do the job.

Where an air cavity between layers of brick is desired, metal ties provide the best means of bridging across the cavity for

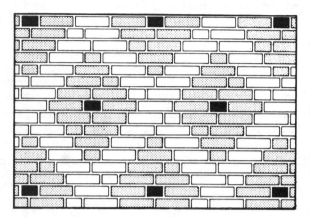

Fig. 4-19. A dovetailed garden-wall design.

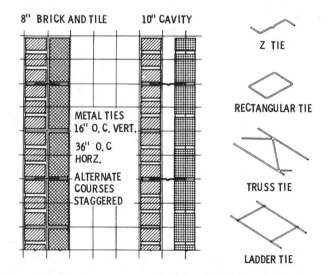

Fig. 4-20. Metal ties bonding together two walls. Various metal ties are also shown.

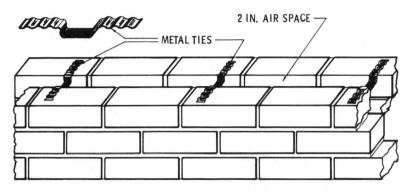

Fig. 4-21. Metal strap ties between two tiers of brick with an air cavity between.

bonding. The center illustration of Fig. 4-20 shows this application with clay tile as the backing wall. Fig. 4-21 illustrates the use of metal ties in an all-brick wall with a cavity between.

While a flat metal tie is shown in Fig. 4-20, most ties are formed of round steel reinforcing rods. The diameter of the rods must be small enough to be fully embedded in the mortar. Since metal is subject to rust, a careful job of mortar work is necessary to keep water seepage to an absolute minimum.

CHAPTER 5

How to Lay Brick

The art of laying brick has not changed much in thousands of years, and happily it is an art well within the capacity of the careful "do-it-yourselfer." Brick itself has not changed a great deal either, though there are some new forms. And, most important, brickwork still looks modern and still has a durability that few other materials can equal. What follows is a variety of techniques to learn how to lay brick.

ESTIMATING NEEDS

Table 5-1 shows the amount of brick and mortar needed for various areas of brick wall surface and at several common thicknesses. This table is based on ½"-thick mortar, which, with the use of standard brick sizes, will result in the 4" modular dimensions. Since actual brick sizes will vary slightly from processor to processor, it will be necessary to adjust the mortar thickness to achieve the modular dimensions.

The dimensional volume of brick to mortar is about 7 to 1. A ⅜″ mortar thickness would mean a 25% reduction in mortar

Table 5-1. Amount of Brick and Mortar Needed for Various Wall Sizes

 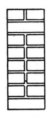 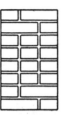 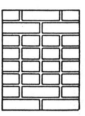

Area of Wall, sq. ft.	4-inch Wall		8-inch Wall		12-inch Wall		16-inch Wall	
	Number of bricks	Cubic feet of mortar	Number of bricks	Cubic feet of mortar	Number of bricks	Cubic feet of mortar	Number of bricks	Cubic feet of mortar
1	6.2	0.075	12.4	0.195	18.5	0.314	24.7	0.433
10	62	1	124	2	185	3½	247	4½
20	124	2	247	4	370	6½	493	9
30	185	2½	370	6	555	9½	740	13
40	247	3½	493	8	740	13	986	17½
50	309	4	617	10	925	16	1,233	22
60	370	5	740	12	1,109	19	1,479	26
70	432	5½	863	14	1,294	22	1,725	31
80	493	6½	986	16	1,479	25	1,972	35
90	555	7	1,109	18	1,664	28	2,218	39
100	617	8	1,233	20	1,849	32	2,465	44
200	1,233	15	2,465	39	3,697	63	4,929	87
300	1,849	23	3,697	59	5,545	94	7,393	130
400	2,465	30	4,929	78	7,393	126	9,857	173
500	3,081	38	6,161	98	9,241	157	12,321	217
600	3,697	46	7,393	117	11,089	189	14,786	260
700	4,313	53	8,625	137	12,937	220	17,250	303
800	4,929	61	9,857	156	14,786	251	19,714	347
900	5,545	68	11,089	175	16,634	283	22,178	390
1,000	6,161	76	12,321	195	18,482	314	24,642	433
2,000	12,321	151	24,642	390	36,963	628	49,284	866
3,000	18,482	227	36,963	584	55,444	942	73,926	1,299
4,000	24,642	302	49,284	779	73,926	1,255	98,567	1,732
5,000	30,803	377	61,605	973	92,407	1,568	123,209	2,165
6,000	36,963	453	73,926	1,168	110,888	1,883	147,851	2,599
7,000	43,124	528	86,247	1,363	129,370	2,197	172,493	3,032
8,000	49,284	604	98,567	1,557	147,851	2,511	197,124	3,465
9,000	55,444	679	110,888	1,752	166,332	2,825	221,776	3,898
10,000	61,605	755	123,209	1,947	184,813	3,139	246,418	4,331

NOTE: Mortar joints are ½″ thick.

volume and about a 3% increase in brick numbers for the same area coverage. A ⅝″ mortar thickness calls for a 25% increase in mortar and 3% decrease in brick numbers. If face brick is used on the first thickness of wall and common brick on other tiers, adjust the number of different bricks accordingly. Remember to allow for extra face bricks, depending on how many face-brick headers are to be laid.

Table 5-2 shows the amount of cement, lime, and sand to use for batches of a little over 1 cu. yd. of mortar. The lower amounts of lime are to be used when portland masonry cement is used. The higher sand ratios are permitted only if the sand is well

Table 5-2. Ratio of Cement, Sand, and Lime for Various Mortar Mixes

Mix by Volume, Cement-Lime-Sand	Cement, Sacks	Lime, lb.	Sand, cu. yd.
1—0.05—2	13.00	26	0.96
1—0.05—3	9.00	18	1.00
1—0.05—4	6.75	44	1.00
1—0.10—2	13.00	52	0.96
1—0.10—3	9.00	36	1.00
1—0.10—4	6.75	27	1.00
1—0.25—2	12.70	127	0.94
1—0.25—3	9.00	90	1.00
1—0.25—4	6.75	67	1.00
1—0.50—2	12.40	250	0.92
1—0.50—3	8.80	175	0.98
1—0.50—4	6.75	135	1.00
1—0.50—5	5.40	110	1.00
1—1—3	8.60	345	0.95
1—1—4	6.60	270	0.98
1—1—5	5.40	210	1.00
1—1—6	4.50	180	1.00
1—1.5—3	8.10	485	0.90
1—1.5—4	6.35	380	0.94
1—1.5—5	5.30	320	0.98
1—1.5—6	4.50	270	1.00
1—1.5—7	3.85	230	1.00
1—1.5—8	3.40	205	1.00
1—2—4	6.10	490	0.90
1—2—5	5.10	410	0.94
1—2—6	4.40	350	0.98
1—2—7	3.85	310	1.00
1—2—8	3.40	270	1.00
1—2—9	3.00	240	1.00

graded. Higher volumes of poorly graded sand can only result in a weak mortar with poor bonding qualities.

SCAFFOLDING

Equally important to the handling of the trowel for reducing worker fatigue and improving production are the placement of the material and the amount of physical movement required to lay brick. The mortar board and supply of brick should be immediately behind the bricklayer so that he has a minimum of steps to take for material. An apprentice should be kept busy supplying the mortar from a mixer and seeing that the brick pile is always supplied.

A brick wall up to about 4 ft. high can be erected with the brick mason standing at ground level. Above 4 ft., it is necessary that he and his material be raised to reduce the amount of reach. Scaffolding for raising the mason and his material takes many forms, from simple to complex. One- or two-man operations on low walls will find a simple wood scaffold sufficient. Contractors with a variety of jobs will make a worthwhile investment in adjustable tubular-metal scaffolding.

Fig. 5-1 shows the dimensions for making a wood scaffold out of homemade or ready-made saw horses. It is important that the top planks be secured with nails to the trestles to allow freedom of movement without worry about loose planks. Mortar board and a supply of brick are also kept on the scaffold. The scaffold structure should be no closer than 3″ from the wall to be sure it does not bear against the wall and push it out of alignment.

Fig. 5-2 shows a typical tubular-metal scaffold structure. It allows for considerable adjustment of the working platform for a minimum amount of bending and reach. The mason's platform does not bear against the brick wall, as it appears to do in the illustration. It is counterbalanced by the weight of the brick and mortar and needs no front support. This type of scaffolding may be used on multistory structures if brick is laid from the inside. At the finish of the first story of the building, a rough floor is placed and the scaffolding is used on the floor for second-story brickwork.

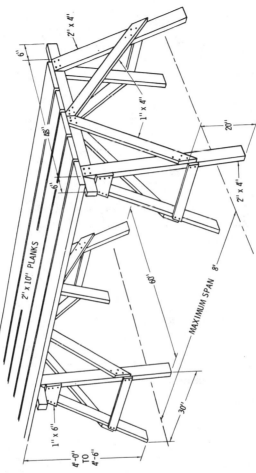

2" x 4"

6'

1" x 4"

20"

48"

6"

2" x 4"

2" x 10" PLANKS

60"

MAXIMUM SPAN 8'

30"

1" x 6"

4'-0"
TO
4'-6"

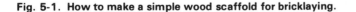

Fig. 5-1. How to make a simple wood scaffold for bricklaying.

FOOTINGS

It is important that the weight of brick walls, especially those which carry a load, be supported by a base that will provide an even distribution of weight. Otherwise, any settling will result in cracks in the mortar.

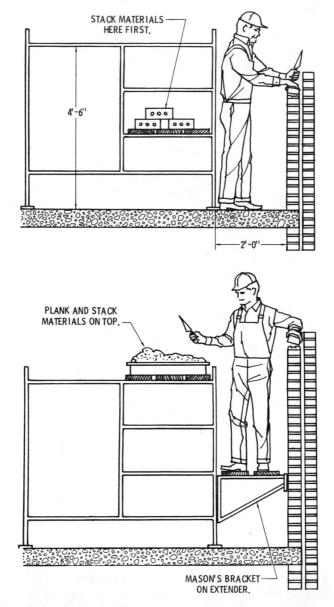

STACK MATERIALS HERE FIRST.

4'-6"

2'-0"

PLANK AND STACK MATERIALS ON TOP.

MASON'S BRACKET ON EXTENDER.

Fig. 5-2. An adjustable tubular-metal scaffold provides greatest flexibility for material placement and height adjustment.

Although some firm, well-packed earth foundations will be sufficient, the usual practice is to pour a concrete foundation in which reinforcing rods have been embedded. The concrete foundation or footing should be twice the width of the wall and have a depth below the frost line. This is particularly important in areas of colder climates, where there is considerable freezing and thawing, which can produce some heaving of the earth. The footing can be poured into a dug trench, with fairly straight sides, with or without forms.

The importance of a perfectly straight and horizontal footing cannot be overemphasized. The footing and the first course of brick will establish the accuracy of a horizontal wall rising course by course above it. After the concrete of the footing has set enough to support the weight of brick above it, usually about seven days, place a line of 1″ thick mortar along it and lay an entire course of brick along the length. Use a spirit level and tap the brick into place to make a perfectly horizontal first course. Thereafter, the taut line method of measuring, described later, will keep the wall going up with horizontal accuracy.

MIXING THE MORTAR

The structural strength of a brick wall depends more on the ingredients and mixing of the mortar and the workmanship in applying it than on the strength of the brick itself. Mortar must have the proper plasticity for easy handling as well as the correct mixture of ingredients for strength over a long period. This was covered in Chapter 2.

A large portion of sand, while economical, will result in weak mortar, unless the sand is well graded. Since well-graded sand costs more than ordinary sand, it may be just as economical to use less sand and more cement to keep the mortar quality good. Mortar must be constantly mixed fresh. If it is more than 2 or 3 hours old, depending on weather, it will begin to take a set and become harder to handle. If the mortar begins to stiffen before that time, it is probably due to water evaporation, and some water may be added (called retempering). If stiffening is due to hydration, retempering will only result in weak mortar.

71

WETTING THE BRICK

Always give the brick the sprinkle test before using. It only takes a minute. Sprinkle a few drops of water on a brick. If it is absorbed by the brick within 1 minute, the brick has too much absorption and should be wetted before using. If left untreated, the brick will absorb water from the mortar, leaving less in the mortar for complete hydration.

If wetting is called for, pour water from a hose on the entire pile of brick until the water runs down the sides. Let the water soak in and the surface water evaporate from the faces of the brick before using them.

PROPER USE OF THE TROWEL

Since the laying of brick is a hand operation and the trowel is constantly in the bricklayer's hand, there is probably no more important operation in laying up a brick wall than the proper use of the trowel. How it is held in the hand and how it is turned as mortar is applied affect efficiency and fatigue.

The student of bricklaying must first learn how to handle the trowel in picking up, throwing, and spreading mortar. The practical way to acquire this knowledge is to practice on the mortar board during lunch periods and before working hours. One should learn how to pick up a trowelful of mortar cleanly and to spread sufficient mortar to lay at least three bricks with one trowelful. Some bricklayers throw, in one operation, enough mortar to lay four or five bricks, depending on the thickness of the joint.

In order to use a trowel properly, it should be held firmly, yet loosely, with the full grasp of the right hand, and applied with the play of the muscles of the arm, wrist, and fingers. Only actual practice can give the various necessary mechanical movements. Lifting a trowelful of mortar from the tub or mortar board up to the courses of brick on the wall and throwing the mortar the length of three or more bricks is done with the muscles of the forearm.

In throwing the mortar, the trowel is turned through a 180° angle (that is, turned upside down) while the trowel is being moved the length of three or more bricks. In order to turn the trowel upside down, evidently it must be held as shown in Fig. 5-3 because the hand, unlike the owl's head, does not work on a pivot and 180° is about the limit it will turn without elevating the elbow. Fig. 5-4 shows how the mortar is "picked up" from the mortar board before throwing it on the brick.

In order to fully understand how the bricklayer places the mortar, the operation is shown in Figs. 5-5 through 5-9. In Fig. 5-5, only the trowel is shown without hand or mortar so that its various positions may be seen as it travels the length of the spread of three bricks and back again to begin the spreading stroke.

Throwing the mortar results in a rounded column of mortar along the central portion of the brick, leaving the outside portions bare, as shown in Fig. 5-6. In order that the brick may have a full bed of mortar to lie on so that the load will be distributed over its entire face, the mortar, after being thrown, should be spread by going over it with the point of the trowel, as shown in Fig. 5-7.

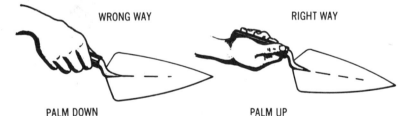

Fig. 5-3. Wrong and right way of holding a trowel. Note the position of the thumb.

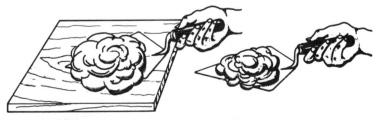

Fig. 5-4. How mortar is taken from the board. The trowel is always held with the thumb up for easier turning.

73

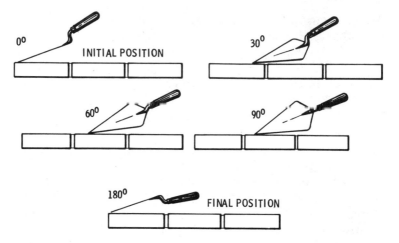

Fig. 5-5. How the trowel is turned during a single stroke of throwing a mortar line.

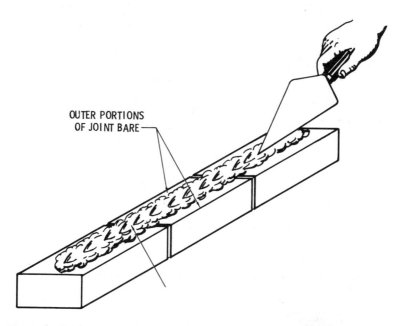

Fig. 5-6. As the trowel is turned, the mortar is spread over the center of a row of three to five bricks.

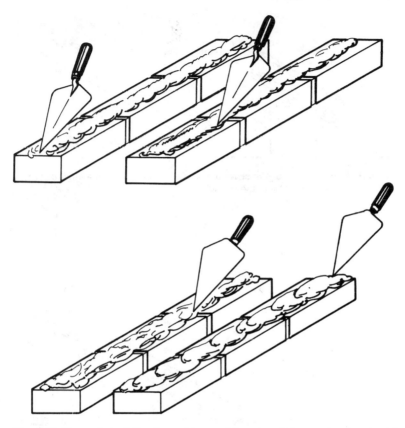

Fig. 5-7. The point of the trowel is used to spread the center furrow of mortar over most of the brick surface.

When the operation of spreading the mortar has been perfectly done, no cutting off is necessary. If, however, too much mortar was thrown or too much pressure was exerted on the trowel in spreading the mortar, some of it will hang over the side of the brick, as shown in Fig. 5-8. In this case it must be cut off so that it will not at any point project over the side of the brick, as shown in Fig. 5-9. In addition to laying a bed of mortar for the brick to lie on, the end of each brick, when laid, must be covered, or "buttered," with mortar so there will be a layer of mortar in the vertical joints as well as in the horizontal joints.

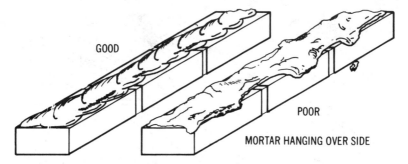

GOOD

POOR

MORTAR HANGING OVER SIDE

Fig. 5-8. Good practice is to spread the mortar over the brick surface without any excessive overhang.

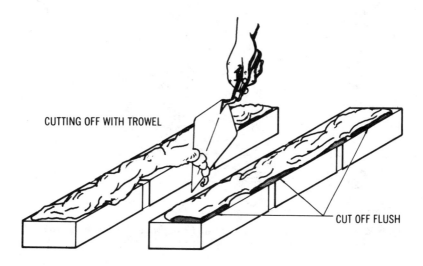

CUTTING OFF WITH TROWEL

CUT OFF FLUSH

Fig. 5-9. If mortar overhang does occur it must be cut off as shown.

LAYING THE BRICK

In lifting a brick from the pile on the ground or scaffold, in order to place it on the bed of mortar, the bricklayer grasps the brick in his left hand, as shown in Fig. 5-10. He butters one end, and in "laying" the brick, first places it on top of the bed of mortar (previously spread) a little in advance (to the right) of its

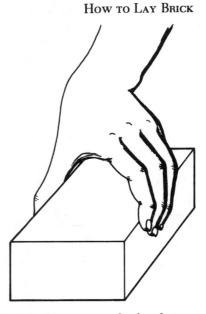

Fig. 5-10. Holding brick ready for placing.

final position, as shown in Fig. 5-11A. He presses the brick into the mortar with a downward slanting motion, as indicated by positions M, S in Fig. 5-11B, so as to press the mortar up into the end joint. During this operation the brick moves from its initial position M, shown in dotted lines M (corresponding to the position shown in Fig. 5-11A), to some intermediate position S, as shown in Fig. 5-11B.

This is the shoving method of bricklaying, and if the mortar is not too stiff and is thrown into the space between the inner and outer courses of brick with some force, it will completely fill the upper parts of the joints not filled by the shoving process. After shoving the brick down and against the mortar in the end joint, it is forced home, or down, until it aligns with the brick previously laid by tapping it either with the blade of the trowel, shown as L (Fig. 5-11C), or with the handle butt of the trowel in position F, shown in dotted lines. During the operation just described, more or less mortar is squeezed out through the face and end joints as shown in Fig. 5-11D. For appearance and to save mortar, it is cut off flush with the trowel. This mortar on the trowel thus cut off should be used for buttering the end of the next brick. It should never be thrown from the trowel back onto the mortar board.

77

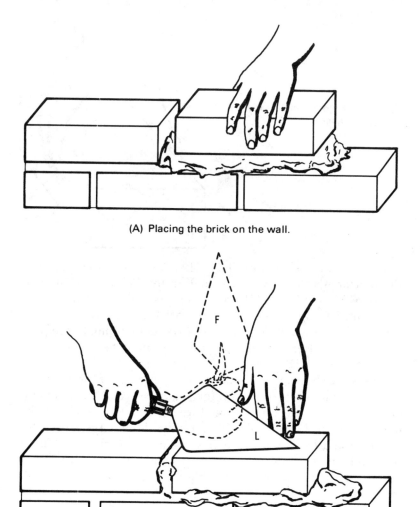

(A) Placing the brick on the wall.

(C) Tapping the brick for proper alignment.

Fig. 5-11. Four steps in

When thrown back onto the mortar board, a large portion may daub up the brick instead of landing on the board, and the operation results in an unnecessary motion each time.

Fig. 5-12 shows the laying of a course of rowlock header brick. In Fig. 5-12A the largest face of the brick has been buttered and is ready to be shoved into position. In Fig. 5-12B a closure is placed as the last brick in the course. Both faces are buttered, and the brick is shoved down into place. If the measurement of total

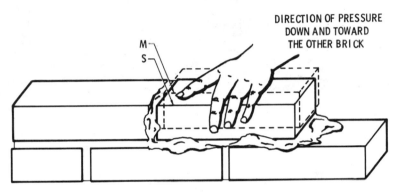

(B) Shoving the brick into position.

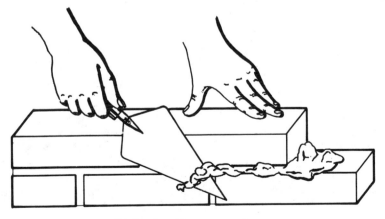

(D) Cutting off excess mortar.

placing a brick into position.

(A) Buttering the wide part of the brick.

(B) Placing the closure or last brick in place.

Fig. 5-12. Rowlock leader brick placement.

length and thickness of mortar has been correct, the closure brick should result in the two facing vertical mortar lines being the same thickness as the others in the wall.

USING THE TAUT LINE

Without a guideline, a true wall surface could not be obtained; some of the bricks would be laid too far out and others too far in. In order to guide the bricklayer so that the brick will lay straight, a taut line secured by pins is used or the equivalent. In order to have supports for the line, a corner of the wall is first built up several courses, and then a lead or support is placed at some point along the course.

The line is made fast around the end or corner, stretched taut, and wound around a brick on the lead, as shown in Fig. 5-13. This is better than using pins or a nail because if the nail pulls loose, it may hit a bricklayer in the eye, resulting in injury to or loss of his eye.

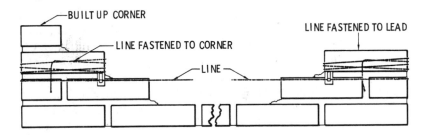

Fig. 5-13. With corners built up, a line is drawn taut to establish a level for each course of brick.

The line should be placed $\frac{1}{32}$″ outside the top edge of the brick and exactly level with it. In order to hold the line at $\frac{1}{32}$″ distance outside the top edge, make two distance pieces out of cardboard or preferably tin, shaped and attached to the line as shown in Fig. 5-14. The reason for this offsetting of the line is that the brick should be laid without touching the line—the $\frac{1}{32}$″ marginal distance being gauged by the eye.

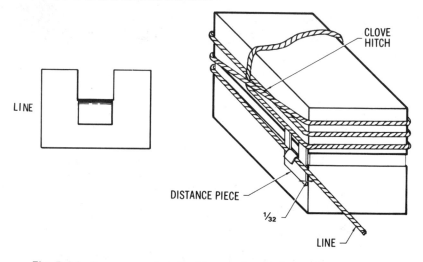

Fig. 5-14. A spacer made to hod the taut line ¹⁄₃₂ " away from the brick face.

Of course, if the bricks are laid so that they touch the line, the latter would be shoved out of place, resulting in irregularities in the wall. Hence, no brick should touch the line. The tendency of inexperienced bricklayers is to "crowd the line," or lay brick *strong* on the line.

The person who works with precision may not be satisfied with just the instruction to set the line level with the top face of the brick. He or she may want to know whether the top or bottom of the line should be level with the top of the brick, especially if it is a thick line. Of course, bricklaying is not a machinist's job, and one is not expected to work with machinist's precision; however, such precision cannot be faulted when it can be used without any extra effort or loss of time. Fig. 5-15 shows the wrong and right ways to set the line with precision.

One should *never touch the line even in applying the mortar or laying the brick*. There are two ways of holding the brick, as shown in Fig. 5-16, so that the line will not be disturbed. It should be understood that even the fingers must not touch the line—otherwise it will be pushed out of place while other workmen are using it as a guide. The method of laying the brick without touching the line is shown in Fig. 5-17. Of course, practice is necessary to do this successfully. The beginner should practice before lay-

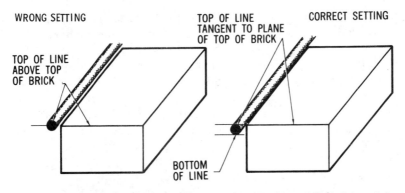

WRONG SETTING

TOP OF LINE
TANGENT TO PLANE
OF TOP OF BRICK

CORRECT SETTING

TOP OF LINE
ABOVE TOP
OF BRICK

BOTTOM
OF LINE

Fig. 5-15. Correctly set the line $\frac{1}{32}$ " out from the edge of the brick and the top surface even with the top of the brick.

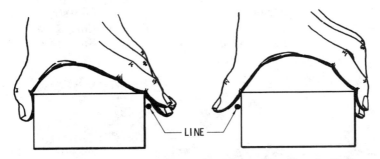

LINE

Fig. 5-16. Placing the brick into position without disturbing the line.

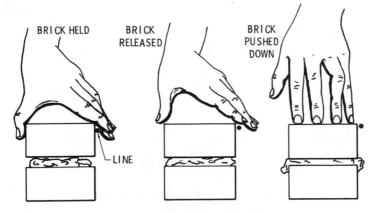

BRICK HELD

BRICK
RELEASED

BRICK
PUSHED
DOWN

LINE

Fig. 5-17. Steps in placing a brick without disturbing the line.

ing to the line so that he will acquire the habit of bringing his thumb and fingers up as the brick goes down near the line.

Fig. 5-18 shows the laying and cutting off of the front tier of a wall. In Fig. 5-18A the end of a brick has been buttered, placed in position and shoved toward the brick already in position to the correct vertical mortar-joint thickness. If the mortar is plastic enough, the pressure of the hand on the brick will align it with the taut string. In Fig. 5-18B a closure brick is placed and pressed down to the level of the line. Fig. 5-18C shows a brickmason cutting off the squeezed-out mortar. With experience this can be done without touching the line.

In Fig. 5-19 a second layer of brick, with a shallow cavity between, is laid and mortar cut off. In Fig. 5-19A the line of mortar is thrown and readied for a few bricks of a course. In Fig. 5-19B the bricks are in place, properly aligned to the string. The only difference between the two series of illustrations is the final cutting off of excess mortar. With a shallow cavity, cutting off

(A) The brick is placed and shoved to the right, aligning it with the string.

Fig. 5-18. Brick placement

84

(B) A closure brick put into place.

(C) Mortar cutoff which will be used to butter the next brick.

and mortar cutoff.

85

(A) Mortar is laid over several bricks.

(B) Bricks in place and aligned.

Fig. 5-19. Laying the rear tier

excess mortar upward is difficult because of the small amount of room. It is more economical to cut off downward and let the excess mortar fall into the cavity, as shown in Fig. 5-19C.

Fig. 5-20 shows a tier of bricks being laid up for the front of a frame home. Because only one tier is used, this is called *brick-veneer construction*. In Fig. 5-20A the taut line is in place and a layer of mortar is thrown across between two halves of a course, with a few bricks in place near the center. Two brick masons are working, one at each half of the course. In Fig. 5-20B one mason works on the short course around the corner. He taps a ½-bat, with header end showing, into place, with the constant use of a spirit level. Note the vertical board temporarily nailed into place. This board has been carefully aligned with a vertical level and serves as the guide for the corner. In Fig. 5-20C the closure brick has been placed and excess mortar has been cut off. The line will now be moved up (⅔ of 4″ for modular brick), checked for horizontal accuracy, and the next course laid.

(C) Mortar cutoff is downward, allowing excess to drop into cavity.

of brick with a cavity between.

(A) With a line in place, mortar is thrown across several bricks.

(B) A ½-bat header is being tapped into place.

Fig. 5-20. Laying up

TOOLING THE JOINTS

Tooling consists of compressing the squeezed-out mortar of the joints back tight into the joints and taking off the excess mortar. The tool should be wider than the joint itself. Jointing tools are available in a number of sizes and shapes. They are generally made of pressed sheet steel or solid tool steel, S-shaped, and are convex, concave, or V-shaped. The convex side is pressed against the mortar.

By pressing the tool against the mortar you will make a concave joint—a common joint, but one of the best. Tooling not only affects appearance, but it also makes the joint watertight, which is the most important function. It helps to compact and fill voids in the mortar. Fig. 5-21 shows concave-joint tooling.

(C) A closure brick has been placed to complete one course.

a brick-veneer wall.

Fig. 5-21. Using a pointing tool to press mortar into the joints.

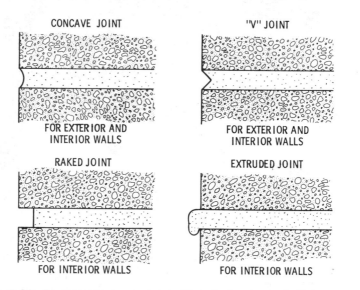

CONCAVE JOINT

FOR EXTERIOR AND
INTERIOR WALLS

"V" JOINT

FOR EXTERIOR AND
INTERIOR WALLS

RAKED JOINT

FOR INTERIOR WALLS

EXTRUDED JOINT

FOR INTERIOR WALLS

Fig. 5-22. Four popular mortar joints. The raked and extruded joints are not recommended for exterior walls in cold climates.

TYPES OF JOINTS

The concave and V-joints are the best for most areas. Fig. 5-22 shows four popular joints. While the raked and the extruded styles are recommended for interior walls only, they may be used outdoors in warm climates where rain and freezing weather are at a minimum. In climates where freezing can take place, it is important that no joint permits water to collect.

In areas where the raked joint can be used, you may find it looks handsome with slump-style brick. The sun casts dramatic shadows on this type of construction.

CHAPTER 6

Wall Types, Thickness, and Anchoring

Brick construction is used for many purposes, but by far the greatest use is in wall construction. Brick features great strength, fire resistance, and good insulation.

Chapter 1 includes information on the compressive strength of brick. Because of this strength, for years multistory structures were made of all brick, with the brick carrying the loads of upper floors and their contents.

Perhaps the best example of all-brick construction is the Monadnock Building in Chicago, Ill., which was built in 1893 and still stands. Six-foot-thick walls at the bottom are tapered to 12″ walls in the top floor. However, since the advent of the skyscraper, steel skeletons carry the loads and brick or other types of walls are used on the outside as well as for interior purposes. As a result, solid brick buildings have been specified only for buildings of three- or four-story heights and less.

A number of years ago, engineers in Europe developed brick-wall designs of only 6″ thickness capable of carrying the loads of buildings up to 16 stories. This design is used in buildings in the United States. Fig. 6-1 shows the design used to provide good longitudinal strength. Fig 6-2 shows the method for achieving good transverse strength. In actual practice, both methods are combined.

TYPES OF BRICK WALLS

There are three basic types of walls in common use. The *veneer* wall is one tier of brick and does not carry a load; it is used only as a facing on frame homes. The *solid* brick wall is from 8″ to 24″ thick, depending on the load it may carry. The *cavity* wall includes an air cavity between the first tier of brick and succeeding tiers of different thicknesses. Cavity walls may consist of brick in all tiers or a combination of brick and hollow clay tile.

VENEER WALLS

A typical home with a brick-veneer facing is of frame construction, using 2″ × 4″ wall studding. Ceiling joists and roof rafters are supported on the sills over the 2″ × 4″ walls. The brick facing provides superior insulation and better appearance. The wide choice of brick textures and colors and the beautiful patterns used in brick overlap make it the most varied of home finishes. Many homes now use one or more interior walls of brick, especially the wall on which a fireplace is located.

There are two generally accepted methods of placing a brick veneer wall against the frame construction of a home. These are shown in Fig. 6-3. The most frequently used method is to space the brick tier away from the sheathing on the frame studs. About a 1″ air space is the usual practice. The brick wall is secured to the frame by corrugated metal tabs. There should be a metal tie for each 2 sq. ft. of wall area. The other method is the use of a paperbacked wire mesh against the studs, instead of sheathing.

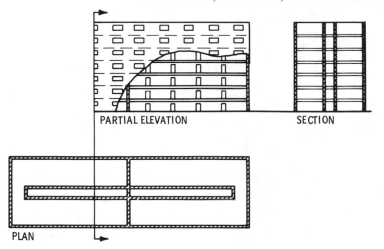

PARTIAL ELEVATION   SECTION

PLAN

Fig. 6-1. Design for longitudinal bearing walls offering high strength. A 6″ wall can bear the load of a 16-story building.

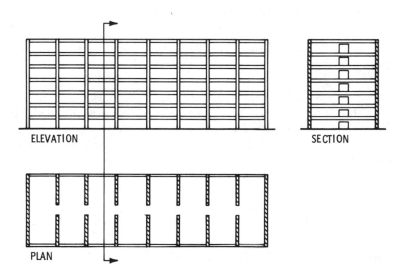

ELEVATION   SECTION

PLAN

Fig. 6-2. Design for high-strength, transverse bearing walls.

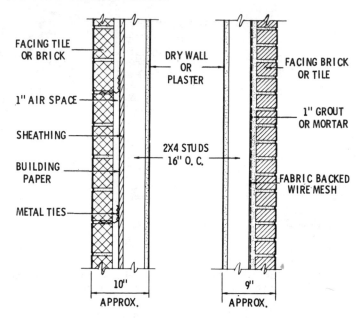

FACING TILE
OR BRICK

DRY WALL
OR
PLASTER

FACING BRICK
OR TILE

1" AIR SPACE

SHEATHING

2X4 STUDS
16" O. C.

1" GROUT
OR MORTAR

BUILDING
PAPER

METAL TIES

FABRIC BACKED
WIRE MESH

10"
APPROX.

9"
APPROX.

Fig. 6-3. Two methods for placing brick veneer on a frame home.

The brick is grouted right up against the wire mesh. A special grout or the regular mortar may be used.

Fig. 6-4 is a cutaway view of brick veneer with the metal ties plainly shown; Fig 6-5 is a side view. The ties should be corrosion-resistant and should be placed with a slight slope downward to the brick to allow any moisture that may be collected to run down toward the front and away from the sheathing.

Extremely important to the use of a brick veneer facing on a frame is the sturdiness of the frame structure, with the least possible give to the pressures of wind or snow load. There is little elasticity to brick, and any pressure from a change in position of the frame can cause serious cracks in the mortar. If there is any suspicion that there may be frame shift or vibration, sheer points should be included in the brick as it is laid up. Fig. 6-6 shows two methods of providing vibration joints if considered necessary. One method is to omit the mortar from one course of brick so that

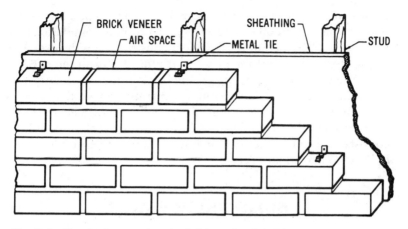

Fig. 6-4. Metal tabs securing the brick to the sheathing.

Fig. 6-5. Side view of the metal-tab holders.

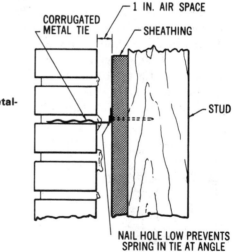

there is no bond between the two courses. To maintain uniform patterns, a mortar line is laid on one course and allowed to set before the next course is laid. Extra ties must be used if this method is employed.

Fig. 6-7 shows two methods for handling the details in laying up a brick veneer wall. One method is for homes with raised

97

floors—those with crawl spaces or basements. The other is for homes constructed on a concrete slab. Alternate foundations are also shown. A feature of brick veneer construction is the excellent barrier formed against the passage of moisture from the outside. Moisture which may get through the brick will flow down the air space instead of passing through the frame construction. This calls for an outlet for the moisture in the form of weep holes at the bottom of the brick. Fig. 6-8 illustrates a typical weep hole. It consists of leaving the mortar out of the vertical joints of the bottom course about every 2 ft.

SOLID BRICK WALLS

Solid brick walls are in common use in one- and two-story homes and two- and three-story apartment buildings. Floor and roof loads are borne principally by the outer walls and to some extent by the inner room walls, usually made of $2'' \times 4''$ framing. Typical wall thicknesses are 8″ and 12″ and typical bonding patterns are those shown in Fig. 6-9. As mentioned, brick is still used extensively for multistory buildings.

The thickness of walls for a multistory building will depend on the load it must bear. This will vary not only by the number of

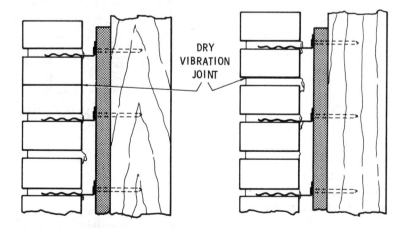

DRY
VIBRATION
JOINT

Fig. 6-6. Two methods of providing vibration joins.

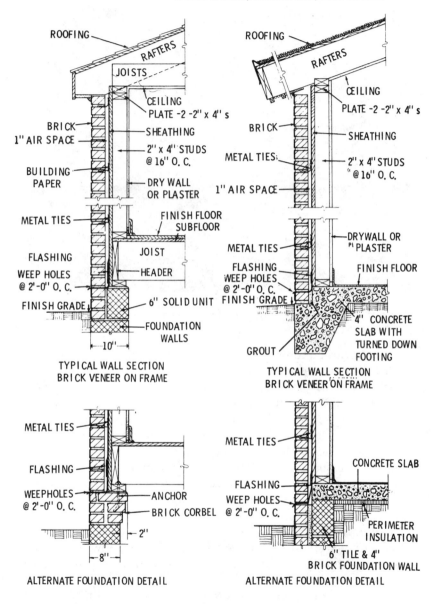

ROOFING

RAFTERS

JOISTS

CEILING

PLATE -2 -2" x 4" s

BRICK

1" AIR SPACE

BUILDING PAPER

METAL TIES

FLASHING

WEEP HOLES @ 2'-0" O. C.

FINISH GRADE

SHEATHING

2" x 4" STUDS @ 16" O. C.

DRY WALL OR PLASTER

FINISH FLOOR SUBFLOOR

JOIST

HEADER

6" SOLID UNIT

FOUNDATION WALLS

10"

TYPICAL WALL SECTION
BRICK VENEER ON FRAME

ROOFING

RAFTERS

CEILING

PLATE -2 -2" x 4" s

BRICK

METAL TIES

1" AIR SPACE

METAL TIES

FLASHING
WEEP HOLES
@ 2'-0" O. C.

FINISH GRADE

GROUT

SHEATHING

2" x 4" STUDS " @ 16" O. C.

DRYWALL OR
ᴾᴵ PLASTER

FINISH FLOOR

4" CONCRETE
SLAB WITH
TURNED DOWN
FOOTING

TYPICAL WALL SECTION
BRICK VENEER ON FRAME

METAL TIES

FLASHING

WEEPHOLES
@ 2'-0" O. C.

ANCHOR

BRICK CORBEL

2"

8"

ALTERNATE FOUNDATION DETAIL

METAL TIES

CONCRETE SLAB

FLASHING
WEEP HOLES
@ 2'-0" O. C.

PERIMETER
INSULATION

6" TILE & 4"
BRICK FOUNDATION WALL

ALTERNATE FOUNDATION DETAIL

Fig. 6-7. Details for using brick veneer.

99

Fig. 6-8. Weep hole at the bottom of a brick-veneer wall.

stories in the building, but by the intended use. For example, industrial buildings may include floors for heavy machinery. The design of walls for heavy loads is quite involved and is determined by engineers and architects. The engineering involved is beyond the scope of this book, but it is important that the apprentice bricklayer knows about the patterns used in laying thick walls since he may be called upon to construct them. What follows, for purposes of demonstration only, are minimum wall specifications. The actual specifications would be more detailed and based on more complicated calculations.

General: The minimum thickness of all masonry bearing or nonbearing walls should be sufficient to resist or withstand all vertical or horizontal loads and comply with the fire resistance requirement of any local code.

Thickness of Bearing Walls: The minimum thickness of masonry bearing walls should be at least 12″ for the uppermost 35 ft. of their height and should be increased 4″ for each successive 35 ft. or fraction thereof, measured downward from the top of the wall.

Exceptions:

1. Stiffened Walls: Where solid masonry bearing walls are stiffened at distances not greater than 12 ft. apart by masonry cross walls or by reinforced concrete floors, they may be of 12″ thickness for the uppermost 70 ft. or fraction thereof.

2. Top-Story Walls: The top-story bearing wall of a building not exceeding 35 ft. in height may be of 8″ thickness, provided it is not over 12 ft. in height and the roof construction imparts no lateral thrust to the walls.

3. One-Story Walls: The walls of a one-story building may be not less than 6″ in thickness, provided the masonry units meet the minimum compressive strength requirement of 2500 psi for the gross area, and that the masonry be laid in Type M, S, or N mortar.

4. Walls of Residence Buildings: In residence buildings not more than three stories in height, walls, other than coursed or rough or random rubble stone walls, may be of 8″ thickness when not over 35 ft. in height. Such walls in one-story residence buildings or private garages may conform to Exception 3.

5. Penthouses and Roof Structures: Masonry walls above roof level, 12 ft. or less in height, enclosing stairways, machinery rooms, shafts, or penthouses may be of 8″ thickness and

8″ FLEMISH BOND   12″ ENGLISH BOND

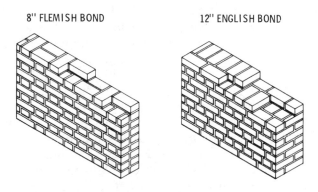

Fig. 6-9. Typical bonding patterns for 8″ and 12″ walls.

may be considered as neither increasing the height nor requiring any increase in the thickness of the wall below.

6. Walls of Plain Concrete: Plain concrete walls may be 2″ less in thickness than required otherwise in this section but not less than 8″, except that they may be 6″ in thickness when meeting the provisions of Exception 3.

7. Cavity Walls: Cavity walls and hollow walls of masonry units should not exceed 35 ft. in height, except that 10″ cavity walls should not exceed 25 ft. in height above the supports of such walls. The facing and backing of cavity walls should each have a nominal thickness of at least 4″, and the cavity should be not less than 2″ (actual) or more than 3″ in width.

8. Composite or Faced Walls: Neither the height of faced (composite) walls nor the distance between lateral supports should exceed that prescribed for the masonry of either of the types forming the facing or the backing.

Thickness of Nonbearing Walls:

1. Exterior Nonbearing Walls: Nonbearing exterior masonry walls may be 4″ less in thickness than required for bearing walls, but the thickness shall not be less than 8″, except where 6″ walls are specifically permitted.

2. Exterior Panel, Apron, or Spandrel Walls: Panel, apron, or spandrel walls that do not exceed 13 ft. in height above their support should not be limited in thickness, provided they meet the first resistive requirements of the code and are so anchored to the structural frame as to ensure adequate lateral support and resistance to wind or other lateral forces.

Fig. 6-10 shows wall thicknesses based on number of brick lengths or widths. Note how the wall thicknesses fit the 4″ module system. Fig. 6-11 shows the various combinations of brick that may be used to make these thicknesses.

Mason contractors for residential dwellings should be acquainted with the building codes in their city. Minimum wall requirements vary somewhat from city to city. The approximate requirements are as follows.

Eight-Inch Walls: It is claimed that a thickness of 8″ for brick walls of the usual home is ample; yet there are numerous cities which do not allow walls under 12″ thick. Some cities allow an 8″ wall for both stories of a two-story house, and many thousands of dwellings have been constructed in these cities with 8″ walls. Further, no city that has adopted the 8″ wall has changed back to the 12″ walls.

THICKNESS MULTIPLE OF 4

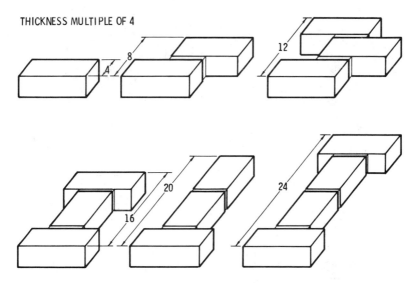

Fig. 6-10. Wall thickness based on 4″ module system of brick sizes.

The discriminating, however, who wish first-class construction will insist on 12″ walls. Some cities require 16″ walls. The brick arrangement for 8″ walls in the various bonds is shown in Fig. 6-12.

Twelve-Inch Walls: For ordinary dwellings, an objection to 12″ walls is the extra space taken up, as compared with 8″ walls; the excess thickness reduces the area of the rooms in the house, which, in cities where land is very valuable, must be taken into consideration. For a house 20 ft. × 30 ft., approximately 31 sq. ft. of area is lost on each story, an area equal to a small bathroom or several good closets.

The extra thickness of the 12″ walls, however, insulates a house

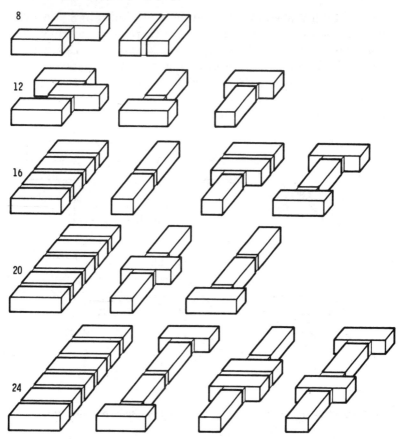

Fig. 6-11. The combinations in which brick may be laid for various wall thicknesses.

better against cold or heat, resulting in a warmer house in the winter and a cooler house in the summer. See Fig. 6-13 for brick arrangements in the various bonds.

Sixteen to Twenty-four-Inch Walls: For heavy duty, as in factory construction where the walls have to carry heavy loads of machinery and are subjected to more or less vibration, the walls may be 16″ to 24″ or more in thickness, depending on conditions.

The arrangement of the brick is more complicated for these thick walls, and the accompanying illustrations have been prepared with progressively extended courses like steps so that the brick arrangement in each course can be clearly seen. These details for 16″ to 24″ walls are shown in Figs. 6-14 to 6-16.

The widest use of brick for interior walls has been in industrial plants. Brick has low heat conduction and high fire resistance, making it ideal for fire walls. Industrial fire walls of brick are partitions and usually are not load bearing. A 4″ thick wall is considered ample, and construction is ½-lap running bond. They are not left free-standing, but are bonded at the ends to adjacent perpendicular walls and to the ceiling above.

Partition walls of brick have many other applications as well. They may be load-bearing, sharing in the load of floors above.

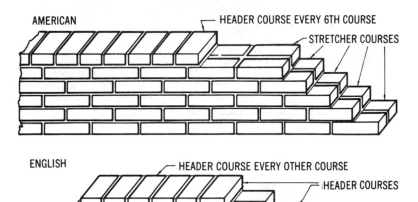

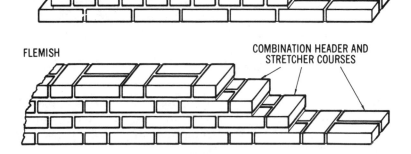

Fig. 6-12. 8″ walls in three popular bonding patterns.

AMERICAN

ENGLISH

HEADER BONDS

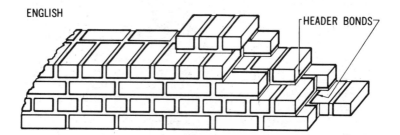

FLEMISH

ALTERNATE BONDS

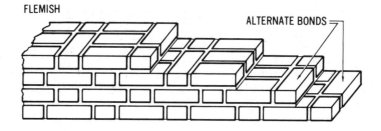

Fig. 6-13. 12″ wall-bonding patterns.

They may be needed for reducing sound transfer from one work area to another. Where interior walls may be subjected to bumping by heavy equipment, the strength of a thick solid brick wall cannot be equalled.

Hollow clay tile is frequently used for interior walls. Tiles may even be combined with solid brick. Although larger in size than brick, clay tile follows the 4″ module size system. (See Chapter 10 for details on structural clay tile.) Brick or clay tile may be left unfinished or may be plastered. Fig. 6-17 illustrates a number of typical interior brick and clay tile walls.

AMERICAN

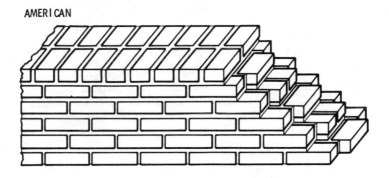

ENGLISH

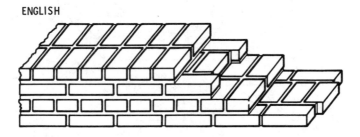

FLEMISH

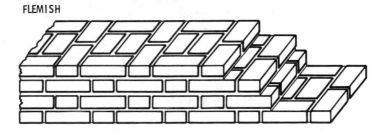

Fig. 6-14. 16″ wall-bonding patterns.

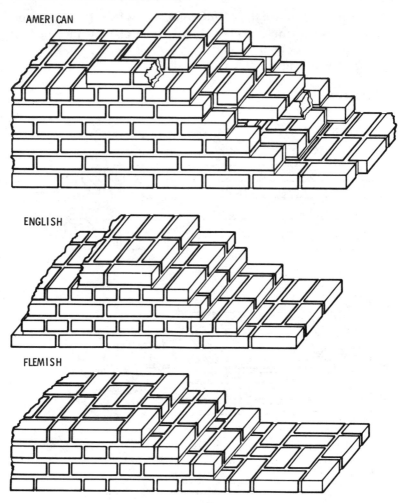

AMERICAN

ENGLISH

FLEMISH

Fig. 6-15. 20" wall-bonding patterns.

CAVITY WALLS

Brick walls are not entirely impervious to the seepage of water. The need to protect against water entrance through a wall becomes important in areas of high wind, coupled with heavy rainfalls. The two maps in Fig. 6-18 show the possible wind velocities and annual rainfall in various areas of the United States. The need to protect against water seepage through brick walls becomes greatest in those areas which combine both high winds and heavy rains, such as along the East and Gulf Coasts and

AMERICAN

ENGLISH

FLEMISH

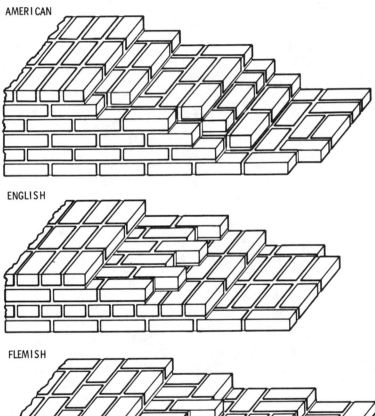

Fig. 6-16. 24" wall-bonding patterns.

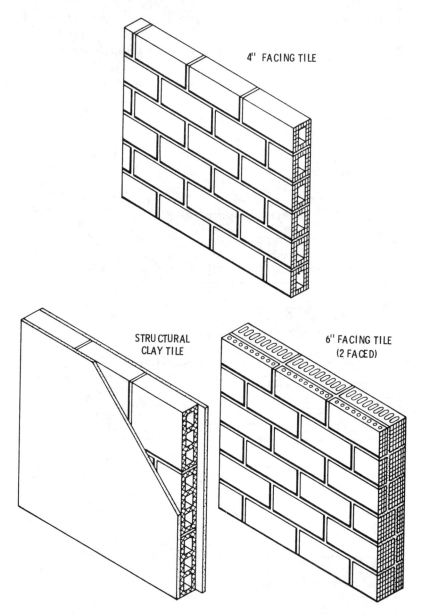

4" FACING TILE

STRUCTURAL
CLAY TILE

6" FACING TILE
(2 FACED)

Fig. 6-17. Typical interior or partition

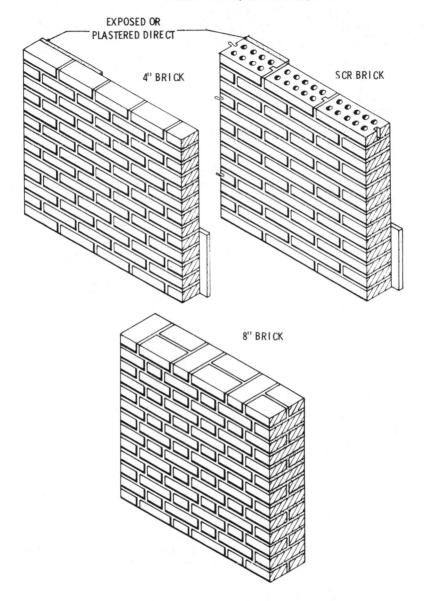

EXPOSED OR
PLASTERED DIRECT

4" BRICK

SCR BRICK

8" BRICK

walls of brick and hollow clay tile.

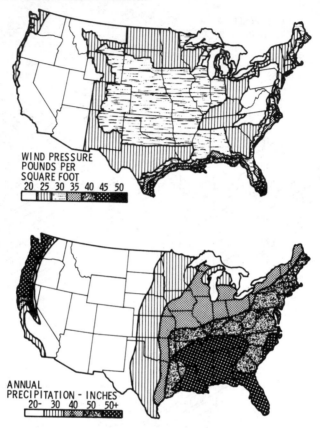

Fig. 6-18. Wind and precipitation maps.

nearly all of Florida. The best answer to the prevention of water seepage is the use of cavity-wall construction.

Cavity, or dual, walls are those in which two adjacent walls are separated by an air space. A cavity wall, therefore, is made up of two walls or tiers of masonry, each nominally 4″ thick with an air space between the walls or tiers for moisture resistance and thermal insulating properties.

Cavity-wall construction is made up entirely of masonry materials, such as: brick on the outside and brick on the inside; brick on the outside and tile or concrete block on the inside; tile on the

112

outside and tile on the inside; or any of the other masonry materials generally used for load-bearing wall construction. In each case there are two walls separated with an air space. Because there are two walls, architects sometimes call cavity-wall construction dual- or barrier-wall masonry.

As illustrated in Fig. 6-19, the two walls are bonded, or tied together, with corrosion-proof, durable, and rigid metal ties. No brick headers are employed, and the stretcher or running bond is generally used. Pattern bonds, such as Flemish bond, can also be used; they are, however, relatively expensive because of the cutting of bricks, resulting in higher labor cost.

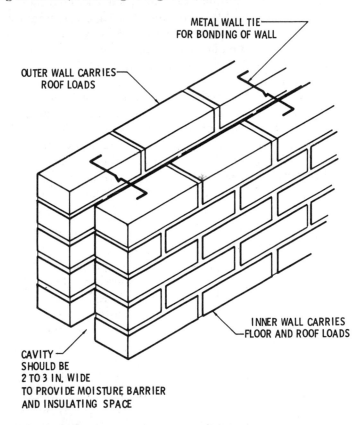

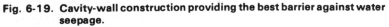

Fig. 6-19. Cavity-wall construction providing the best barrier against water seepage.

Cavity-wall construction has been used extensively for both commercial and residential construction in continental Europe for over a century and more recently in the United States, where it has had an ever-increasing popularity.

Safety Considerations

Cavity walls are generally accepted throughout the United States by local and national codes. These codes as generally recognized are for general safety purposes. Cavity walls for residences and multistory buildings are generally constructed with a 10″ overall thickness. Institutional buildings, such as churches and schools of two-story, load-bearing construction, are generally 14″ in overall thickness. Regardless of the thickness of the cavity wall, it is limited to a height of 35 ft. and must be supported at right angles to the wall face at intervals not exceeding 14 times the nominal wall thickness.

The 10″ cavity walls are limited to not more than 25 ft. in height and must be similarly supported laterally. This lateral support can be from the roof, floors, or partitions. Many buildings of multistory design have been constructed of curtain walls having cavity-wall construction. By resting on spandrels or shelf angles, they do not exceed the height limitations. Their interior treatment has varied from exposed masonry wall to plaster over lath and furring.

Dry-Wall Construction

The cavity wall was originally erected to provide a physical separation between the outside and the inner wall. It was determined many years ago that walls that permitted water penetration were generally of a character in which the mortar did not bond properly to the brick, and as a consequence wind-driven rains would penetrate through the hairline cracks created by this lack of bond. Moisture appearing on the interior surface of the wall frequently caused a disintegration of wall finishes such as plaster or wood paneling.

Building with cavity walls prevents moisture from going through the wall. As shown in Fig. 6-20, any moisture that may gain entrance through the exterior tier of masonry will run down

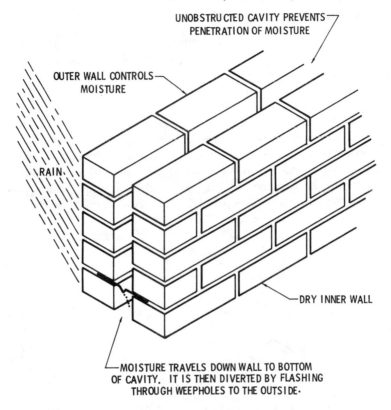

UNOBSTRUCTED CAVITY PREVENTS PENETRATION OF MOISTURE

OUTER WALL CONTROLS MOISTURE

RAIN

DRY INNER WALL

MOISTURE TRAVELS DOWN WALL TO BOTTOM OF CAVITY. IT IS THEN DIVERTED BY FLASHING THROUGH WEEPHOLES TO THE OUTSIDE.

Fig. 6-20. How cavity walls prevent the penetration of moisture from the outer wall through to the inner wall.

the inner side or cavity to the bottom of the cavity, where it drains to the outside through weep holes provided for that purpose. Drops of water that may happen to reach the metal wall ties will drop off before they reach the inner wall if the wall tie has the drop loop. Thus the inner wall will generally remain dry. As in solid wall construction, however, the heads of windows, ventilating ducts, and other wall openings should be flashed. It is also very important to keep the cavity clean and clear of any solid material that may act as a bridge for the passage of moisture from the outer to the inner wall since a clogged cavity or one filled with mortar does not perform well or efficiently.

115

Weep Holes

Weep holes, as previously noted, are designed so that any moisture that may go through the outer wall will run down the inner side of the outer wall, as shown in Fig. 6-20. To drain off this moisture, weep holes are used, as shown in Fig. 6-21. As noted in the illustration, weep holes are located in the outer wall in the vertical joints of the bottom course, preferably two courses above grade, and are spaced from three to five bricks apart.

Weep holes are created by various means. The simplest is the omission of mortar from the vertical joint at predesigned intervals in the bottom course, or, preferably, not less than two courses above grade. Weep holes may also be made with ⅜″

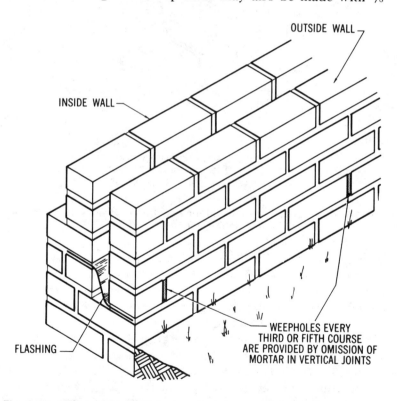

OUTSIDE WALL

INSIDE WALL

WEEPHOLES EVERY
THIRD OR FIFTH COURSE
ARE PROVIDED BY OMISSION OF
MORTAR IN VERTICAL JOINTS

FLASHING

Fig. 6-21. Weep holes placed at various intervals allowing moisture to escape from the wall cavity.

oiled steel rods, pipe, or short lengths of sash cord or rubber hose, which are removed when the mortar sets up. Frequently, a piece of plywood the size of the joint will be put in the vertical joint, and after the mortar sets up, the plywood pieces are removed. At a later date, a 2½" × 2½" square piece of copper screening is wrapped around the piece of plywood and inserted in the weep hole as precaution against the entrance of bugs into the cavity. The plywood is then removed, leaving the screen in place. When the holes or sash cord are used, they should extend up into the cavity and project outward from the base of the exterior wall for easy removal. This ensures a clear and clean weep hole.

Weep holes are necessary for proper drainage to keep the bottom of the cavity dry. Dirt and gravel used in landscaping should not be allowed to pile up higher than the bottom of the cavity. Such material will block the weep holes, and moisture may back up in the cavity so that it gets higher than the flashing and will then penetrate the inner tier of masonry.

Width of Cavity

The width of the cavity may vary considerably. The cavity in this type of wall construction is recommended to be not less than 2" or more than 3" wide, as shown in Fig. 6-22. The reason for this is that a cavity of less than 2" in width permits mortar to bridge the cavity by dripping on the wall ties as it falls down the wall.

A limitation of 3" in width on the cavity has been determined by tests made on the effective width in the cavity. Thus, it has been found that when the width of the cavity approaches 4", the width becomes great enough to cause air coming in from the weep holes to go up the inner side of the outer wall and come down the outer side of the inner wall, thereby defeating a purpose of the construction. The foregoing action is created by an eddying current of air, which is highly undesirable in cavity-wall construction. The cavities between walls may also be used for the installation of heating and air-conditioning ducts, as well as pipes, as shown in Fig. 6-23.

Fig. 6-24 shows several methods of cavity-wall construction in addition to the cavity between two tiers (sometimes called

117

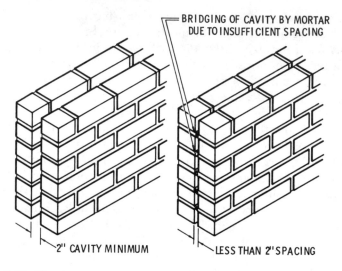

Fig. 6-22. The right and wrong widths for cavity walls.

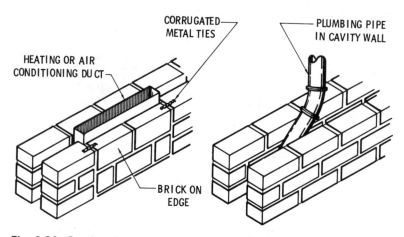

Fig. 6-23. Service pipes and air ducts may be installed in air cavity.

wythes) of brick. Fig. 6-24A shows the brick veneer discussed before. While the cavity is small, it is sufficient for water protection. In Fig. 6-24B, SCR brick is used, a type developed by the Structural Clay Products Institute, now the Brick Institute of

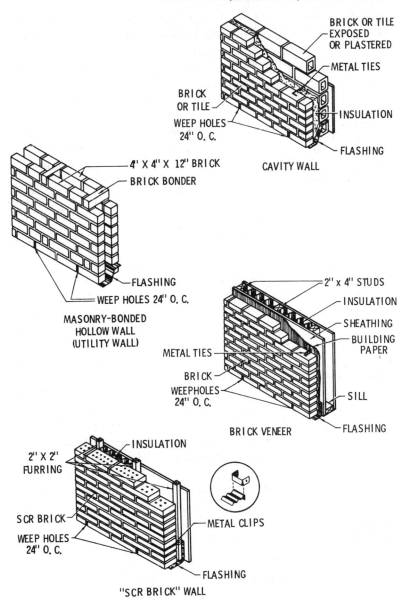

BRICK OR TILE
EXPOSED
OR PLASTERED

METAL TIES

BRICK
OR TILE

INSULATION

WEEP HOLES
24" O. C.

FLASHING

CAVITY WALL

4' X 4" X 12" BRICK

BRICK BONDER

FLASHING

WEEP HOLES 24" O. C.

MASONRY-BONDED
HOLLOW WALL
(UTILITY WALL)

2" x 4" STUDS

INSULATION

SHEATHING

BUILDING
PAPER

METAL TIES

BRICK

WEEPHOLES
24" O. C.

SILL

FLASHING

BRICK VENEER

INSULATION

2" X 2"
FURRING

METAL CLIPS

SCR BRICK

WEEP HOLES
24" O. C.

FLASHING

"SCR BRICK" WALL

Fig. 6-24. Four alternate methods of cavity-wall construction.

119

America. Its construction and width (6" instead of the usual 4") make it especially suited to prevent or reduce water penetration. The cavity wall shown in Fig. 6-24C may include a tile back tier, as shown, or solid brick. The cavity may be left empty (an air cavity) or filled with water-resistant vermiculite or perlite for added insulation. The cavity wall is the best insulator against water.

While bonding of the outer tier to the inner tier is usually done by metal straps (Fig. 6-24D), bonding may be accomplished by using header brick. The tiers are rowlock laid, with bricks on edge. The reduced tier thickness leaves a cavity across which a header brick can reach.

METAL TIES

Metal ties are used in cavity-wall construction to bond the walls and to furnish the necessary rigidity between the tiers. Fig. 6-25 shows the various types of wall ties that have been used in cavity-wall construction and their dimensions. Metal ties should be corrosion proof, rigid, and durable, and not less than $\frac{1}{16}$" in diameter. The most frequently used is the Z-bar, in which the wire is bent to provide a hook about 2" in length for embedment in the horizontal mortar joint of the inner and outer walls.

Building codes usually call for a maximum spacing of ties as one tie for each 3 sq. ft. of wall area. In standard brickwork the spacing accordingly will be about 24" on horizontal centers and each sixth course vertically. Mortar is spread on each wall section before the ties are placed to provide a bond between the tie and the brick.

Regardless of the type of wall tie used, caution should be observed against placement with a pitch toward the inner wall. Wall ties must be placed within 12" of all wall openings and at the bottom of joists or slabs that rest on the wall.

EXPANSION JOINTS

No building is a perfectly rigid structure. There is always some movement, whether it is due to some settling, heavy traffic

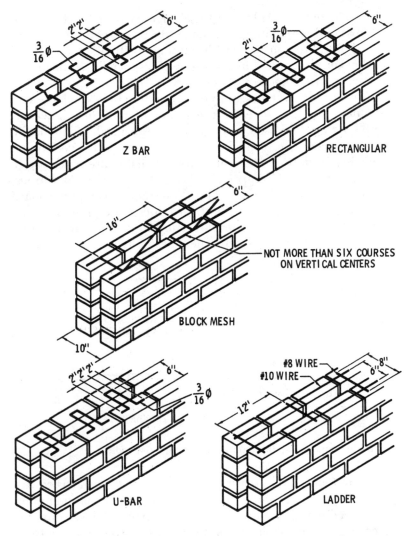

Fig. 6-25. Popular reinforcing rods used for bonding cavity walls.

nearby, possible earth tremors, or other causes. One movement factor no building can avoid is that due to expansion and contraction resulting from temperature and moisture changes.

It takes a 100°F change in temperature to expand a 100-ft.-long

121

brick wall only ⅜″. As little as the change seems to be, it does not take much of a temperature change to cause cracks in masonry walls. For this reason, large walls must not be bonded into a rigid unit but must include joints in the wall, especially vertically, to permit expansion and contraction.

An expansion joint is a complete separation between large sections of walls. To prevent the entrance of moisture, however, some filler must be used in the joint and the installation must be correctly made. Preformed copper sheeting with overlapping seams has been used for years. Now available are other types, such as premolded compressible and elastic fillers of rubber, neoprene, and other plastics. The old system of using fiberboard is not recommended since fiberboard does not compress easily, and once compressed it does not return to its original size. Fig. 6-26 shows some of the acceptable types of expansion joints.

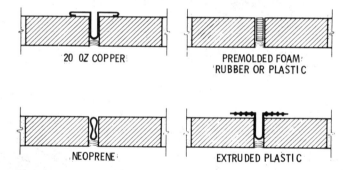

Fig. 6-26. Expansion-joint material in common use today.

Fig. 6-27 shows the treatment of a number of wall structures. Note that metal ties should be used around the expansion joint. The main thing for the brickmason to keep in mind is the importance of a good job in the installation of expansion joints. If anything in the joint prevents free movement, the effectiveness of the expansion joint is completely ruined. Mortar or pieces of broken brick must not be allowed to fall into the joints. The full length of the joint must be kept free of dirt or any type of rubble. If it is necessary to use ties across the joint, they must be of the flexible type, and the ties must not be bridged with mortar.

FOUNDATIONS

A necessity for laying up a wall with horizontal courses and to provide long life with a minimum of settling is to begin on a good foundation. While brick may be used as the foundation, modern practice is to pour a concrete footing.

The depth of the foundation must be at least below the frost line, and the width should be twice the width of the wall. A trench dug with straight sides into which reinforcing steel rods have been placed is often a sufficient form. Wood forms may be placed if the earth is not solid enough to develop straight sides. It is important that the top surface of the foundation be as horizontal as possible. Adjustment can be made at the time of placing a mortar line and the first course of brick.

Detailed information on concrete foundations appears in another volume. The following is a summary of that information, included here for convenience.

Foundation Materials—Foundations, owing to the availability of material, are usually built of concrete. The materials necessary for making concrete are cement, sand, aggregate, water, and, in some instances, reinforcement. It has, however, become customary to refer to concrete as having only three ingredients, namely, cement, sand, and aggregate, the combination of which is expressed as a mixture by volume in the order referred to. Thus, for example, a concrete mixture referred to as 1:2:4 actually means that the mixture contains 1 part of *cement*, 2 parts of *sand*, and 4 parts of *aggregate*, each proportioned by volume.

The general practice of omitting water from the ratio does not necessarily mean that the amount of water used in the the mixture is less important; it is omitted partly to simplify the formula and also because the amount of water used involves a consideration of both the degree of exposure and strength requirements of the complete structure.

Cement Ratio—Table 6-1 gives recommended water-cement ratios on the basis of a definite minimum curing condition for concrete to meet different degrees of exposure in different classes of structures.

In determining the proportions of materials, it is desirable to arrive at those proportions which will give the most economical

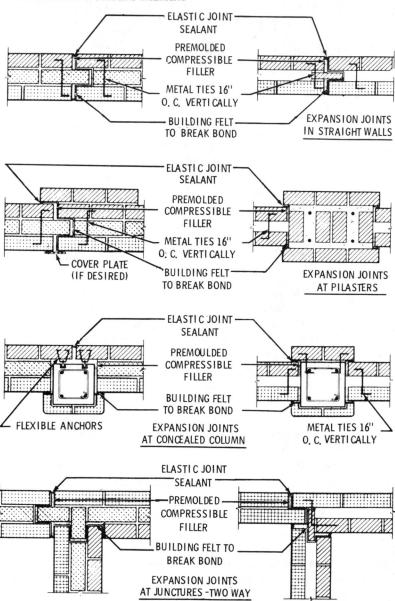

ELASTIC JOINT SEALANT

PREMOLDED COMPRESSIBLE FILLER

METAL TIES 16" O. C. VERTICALLY

BUILDING FELT TO BREAK BOND

EXPANSION JOINTS IN STRAIGHT WALLS

ELASTIC JOINT SEALANT

PREMOLDED COMPRESSIBLE FILLER

METAL TIES 16" O. C. VERTICALLY

COVER PLATE (IF DESIRED)

BUILDING FELT TO BREAK BOND

EXPANSION JOINTS AT PILASTERS

ELASTIC JOINT SEALANT

PREMOULDED COMPRESSIBLE FILLER

BUILDING FELT TO BREAK BOND

FLEXIBLE ANCHORS

EXPANSION JOINTS AT CONCEALED COLUMN

METAL TIES 16" O. C. VERTICALLY

ELASTIC JOINT SEALANT

PREMOLDED COMPRESSIBLE FILLER

BUILDING FELT TO BREAK BOND

EXPANSION JOINTS AT JUNCTURES - TWO WAY

Fig. 6-27. Examples of placement of

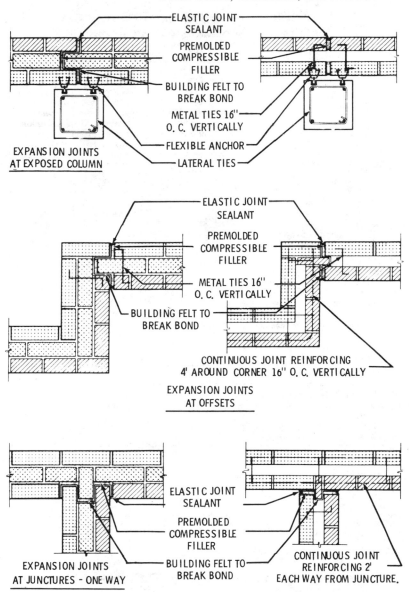

ELASTIC JOINT SEALANT

PREMOLDED COMPRESSIBLE FILLER

BUILDING FELT TO BREAK BOND

METAL TIES 16" O. C. VERTICALLY

FLEXIBLE ANCHOR

LATERAL TIES

EXPANSION JOINTS AT EXPOSED COLUMN

ELASTIC JOINT SEALANT

PREMOLDED COMPRESSIBLE FILLER

METAL TIES 16" O. C. VERTICALLY

BUILDING FELT TO BREAK BOND

CONTINUOUS JOINT REINFORCING 4' AROUND CORNER 16" O. C. VERTICALLY

EXPANSION JOINTS AT OFFSETS

ELASTIC JOINT SEALANT

PREMOLDED COMPRESSIBLE FILLER

BUILDING FELT TO BREAK BOND

EXPANSION JOINTS AT JUNCTURES - ONE WAY

CONTINUOUS JOINT REINFORCING 2' EACH WAY FROM JUNCTURE.

expansion joints in various wall structures.

Table 6-1. Recommended Water-Cement Ratio
for Different Applications and Conditions

Exposure	Water-Cement Ratio, U.S. gal. per sack*		
	Reinforced piles, thin walls, light structural members, exterior columns, and beams in buildings	Reinforced reservoirs, water tanks, pressure pipes, sewers, canal linings, dams of thin sections	Heavy walls, piers, foundations, dams of heavy sections
Extreme: 1. In severe climates such as in northern U.S., exposure to rain, snow, drying, freezing, and thawing, as at the water line in hydraulic structures. 2. Exposure to sea and strong sulphate waters in both severe and moderate climates.	5½	5½	6
Severe: 3. In severe climates such as in northern U.S., exposure to rain, snow, freezing, and thawing, but not continuously in contact with water. 4. In moderate climates such as in southern U.S., exposure to alternate wetting and drying, as at water line in hydraulic structures.	6	6	6¾
Moderate: 5. In climates such as in southern U.S., exposure to ordinary weather, but not continuously in contact with water. 6. Concrete completely submerged but protected from freezing.	6¾	6	7½
Protected: 7. Ordinary enclosed structural members; concrete below the ground and not subject to action of corrosive groundwaters or freezing and thawing.	7½	6	8¼

*Surface water or moisture carried by the aggregate must be included as part of the mixing water.

results consistent with proper placing. The relative proportions of fine and coarse aggregates and the total amount of aggregate that can be used with fixed amounts of cement and water will depend, not only on the consistency of concrete required, but also on the grading of each aggregate. A combination of aggregates made up largely of coarse particles presents less total surface to be coated with cement paste than aggregate of fine particles and is therefore more economical. For this reason it is desirable to use the lowest proportion of fine aggregate which will properly fill the "void" spaces in the coarse aggregate.

Aggregates that are graded so that they contain many sizes are more economical than aggregates in which one or two sizes predominate because the former contain fewer voids. The small particles fill the spaces between the larger particles, which otherwise must be filled with cement paste. A properly proportioned combination of well-graded fine and coarse aggregates contains all sizes between the smallest and the largest without an excessive amount of any one size. The best grading, however, is not necessarily one consisting of equal amounts of the various sizes because such a grading is seldom practicable. Satisfactory mixtures can usually be obtained with the commercial aggregates by proper combination of fine and coarse aggregates.

Increasing the proportion of coarse aggregate up to a certain point reduces the cement factor. Beyond this point the saving in cement is very slight, while the deficiency in mortar increases the labor cost of placing and finishing. Because coarser gradings are more economical, there has been a tendency to use mixtures that are undersanded and harsh. Harshness has been the principal cause for overwet mixtures, resulting almost invariably in honeycombinng in the finished work. While increasing the proportion of fine materials makes for smoother working mixes, excessive proportions of fine materials present greater surface areas to be coated and more voids to be filled with cement paste. Under such conditions, the total amount of aggregate which can be used with fixed amounts of cement and water is greatly reduced.

The total amount of aggregate that can be used with given amounts of cement and water will depend on the consistency required by the conditions of the job. A stiffer mix permits more aggregates to be crowded into the cement paste and thus gives a larger volume of concrete. Stiffer mixes costs less for materials than the more fluid mixes, but the cost of handling and placing increases when excessively dry mixes are used. On the other hand, mixes that are overwet require high cement factors and cannot be placed without segregation of the materials. Such mixes are uneconomical in material and are seldom required for the conditions of placing. In many instances, where correct proportions of sand are used, it will be found practicable to use somewhat stiffer mixtures than has been the practice in the past, without adding materially to the cost of handling or placing.

Because of the restrictions imposed by limiting the amount of water for each sack of cement, experienced workers can generally be depended upon to obtain a proper balance between the various factors, with the result that the concrete will be neither harsh or honeycombed, on the one hand, nor porous and over-wet, on the other.

One of the important advantages to the contractor of the water-cement ratio method is that the materials may be proportioned to facilitate handling and placing, thereby reducing the cost of these items. With some latitude in the matter of workability and proportions, he will be quick to select those mixes which give him the necessary workability at the lowest cost. At the same time, such a mixture will thoroughly fill the forms and reduce the cost of patching honeycomb spots to a minimum. Where the surfaces are to be given a special treatment, the process is invariably made easier.

The quantities of materials in a concrete mixture may be determined accurately by making use of the fact that the volume of concrete produced by any combination of materials, as long as the concrete is plastic, is equal to the sum of the absolute volume of the cement plus the absolute volume of the aggregate plus the volume of water. The absolute volume of a loose material is the actual total volume of solid matter in all the particles. This can be computed from the weight per unit volume and the apparent specific gravity as follows:

$$\text{Absolute vol.} = \frac{\text{unit weight}}{\text{apparent specific gravity} \times \text{unit wt. of water}}$$
$$\text{(62.5 lb. per cu. ft.)}$$

in which the unit weight is based on surface dry aggregate.

The method can best be illustrated by an example. Suppose the concrete batch consists of one sack of cement (94 lb.), 2.2 cu. ft.

of dry fine aggregate weighting 110 lb. per cu. ft., and 3.6 cu. ft. of dry coarse aggregate weighing 100 lb. per cu. ft., which is to be mixed with a water-cement ratio of 7 gallons per sack. The apparent specific gravity of the cement is usually about 3.1, and that of the more common aggregates is about 2.65. The volume of concrete produced by the above mix is calculated as follows:

$$\text{Cement} = 1 \text{ cu. ft. at } \frac{94}{3.1 \times 62.5} = 0.49 \text{ cu. ft. abs. vol.}$$

$$\frac{\text{Fine}}{\text{Aggregate}} = 2.2 \text{ cu. ft. at } \frac{110}{2.65 \times 62.5} = 1.46 \text{ cu. ft. abs. vol.}$$

$$\frac{\text{Coarse}}{\text{Aggregate}} = 3.6 \text{ cu. ft. at } \frac{100}{2.65 \times 62.5} = 2.18 \text{ cu. ft. abs. vol.}$$

$$\frac{\text{Volume of}}{\text{Water}} = \frac{7.0}{7.5} = 0.93 \text{ cu. ft. abs. vol.}$$

Total Volume of Concrete Produced = 5.06 cu. ft.

Thus one sack of cement produces 5.06 cu. ft., neglecting absorption or losses in manipulation. The cement required for 1 cu. yd. of concrete is, therefore,

$$\frac{27}{5.06} = 5.34 \text{ sacks}$$

The quantities of fine and coarse aggregate required can be found from a simple computation based on the number of cubic feet used with each sack of cement. Thus, for fine aggregate

$$\frac{5.34 \times 2.2}{27} = 0.43 \text{ cu. yd.}$$

For the coarse aggregate

$$\frac{5.34 \times 3.6}{7} = 0.71 \text{ cu. yd.}$$

Placing of Concrete—No element in the whole cycle of concrete production requires more care than the final operation of placing concrete at the ultimate point of deposit. Before placing concrete, all debris and foreign matter should be removed from the places to be occupied by the concrete, and the forms, if used, should be thoroughly wetted or oiled. Temporary openings should be provided where necessary to facilitate cleaning and inspection immediately before depositing concrete. These should be placed so that excess water used in flushing the forms may be drained away.

Prevention of Segregation—With a well-designed mixture delivered with proper consistency and without segregation, placing of concrete is simplified; but even in this case care must be exercized to further prevent segregation and to see that the material flows properly into corners and angles of forms and around the reinforcement.

Constant supervision is essential to ensure such complete filling of the form and to prevent the rather common practice of depositing continuously at one point, allowing the material to flow to distant points.

Flowing over long distances will cause segregation, especially of the water and cement from the rest of the mass. An excessive amount of tamping or puddling in the forms will also cause the material to separate. When the concrete is properly proportioned, the entrained air will escape and the mass will be thoroughly consolidated with very little puddling. Light spading of the concrete next to the forms will prevent honeycombing and make surface finishing easier.

REINFORCED BRICK MASONRY

Reinforced brick masonry is sometimes called by its initials, RBM. Brick has tremendous compressive strength (as in the case of concrete), but the tensile or lateral strength depends on the bond between bricks. In areas where heavy winds or earth tremors are frequent, it may be necessary to increase the bond between bricks for greater lateral strength. This is done with steel reinforcing rods.

The actual number, size, and placement of the rods becomes an engineering problem. It is important for the brick mason to know how to install them to ensure best bond of the rod to the brick structure. For example, it is important that a full layer of mortar be placed on the brick faces holding the rods. The rods must be imbedded in as much mortar as possible. Furthermore, there must be mortar on all sides of the rods—the rods must be suspended in the center of the mortar thickness. The mortar bed must be at least ⅛″ thicker than the rods so that there is a ⅟₁₆″ thickness of mortar on each side of the rods. Where large-diameter rods are used, this may call for a thicker mortar joint. All joints must have the greater thickness to maintain a uniform pattern.

Fig. 6-28 is an example of steel rods laid the length of an 8″ wall. A transverse rod bridging the two tiers of a cavity wall is

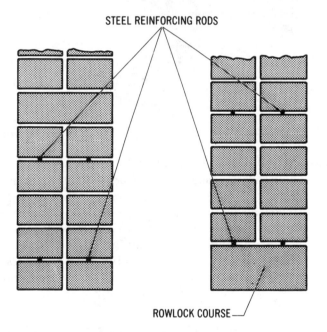

STEEL REINFORCING RODS

ROWLOCK COURSE

Fig. 6-28. Horizontal rods for extra bonding are shown in this end view.

shown in Fig. 6-29. One of the most important applications of reinforcing rods in brick construction is in the erection of columns and pillars made of brick; Fig. 6-30 is an example.

It is common practice to include horizontal reinforcing rods in the concrete poured as a footing for brick walls. Often this

Fig. 6-29. Transverse rods used to bond tiers of brick in a cavity wall.

includes vertical-rod members placed every few feet along the footing. These rods are gauged to protrude into the open space of a cavity brick wall. Concrete is poured into the cavity for a couple of feet in height to bond to the vertical rods. This results in a solid structural mass between the concrete footing and the brick wall. The remaining height may leave the air space in the cavity for bleeding water, as explained before. Weep holes must be included, of course. Fig. 6-31 shows a mason pouring concrete into the cavity using vertical-rod reinforcing.

USE OF ANCHORS

The metal ties for securing one tier of brick to another and the reinforcing rods mentioned above are called *anchors*. However, special heavy-duty forms of anchors are used:

1. To reinforce corners of brickwork.
2. To tie joists and roof plates to the brickwork.

The anchors are made in many shapes to meet the requirements of the service for which they are needed.

Anchoring Walls at Angles

An important feature in brickwork is that the walls should be anchored where they meet at corners; that is, the front and rear walls should be securely anchored as well as bonded to the side, partition, or partition walls. Fig. 6-32 shows some forms of the rods commonly used.

The provision for tying consists of an anchor placed at the center of a 4″ recess or blocking. The T or pin anchor should be built into the center of the recess, which should occur every 13 courses. The anchor should project so as to give not less than 8″ of holding on the wall to be tied. These anchors should never be

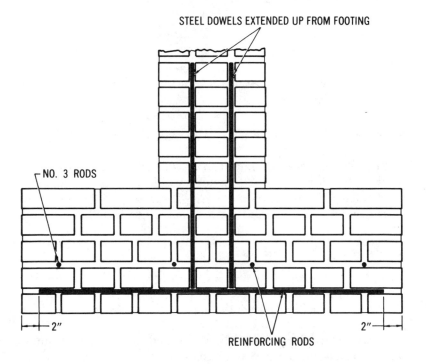

Fig. 6-30. A column of brick using vertical reinforcing rods.

133

Fig. 6-31. Vertical rods extending from the footing tie the bricks to the footing.

omitted when one wall is coursed up before the wall to be tied is built.

Fig. 6-33 shows an anchor in an 8″ wall. It will secure this wall to the adjoining one when the next wall is laid up. In Fig. 6-34, anchors in a 12″ wall are shown. An intersecting wall is tied to an outside wall with a nut and washer anchor in Fig. 6-35.

Anchoring Floor Joints

In brickwork the courses can easily be adjusted so that the courses supporting joists will be at the exact height required. No "shims" or blocking under the joists are needed or should be allowed.

Joists and timbers should be set directly on the brick unless their bearing surface is so small that they transmit a load greater than the safe bearing capacity of the wall, which occurs very

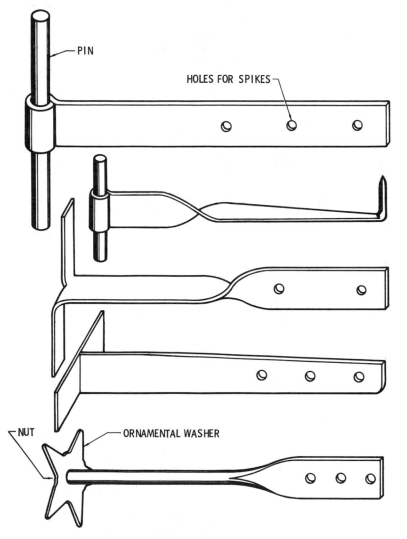

Fig. 6-32. Various forms of heavy anchor rods used in brickwork.

seldom but which would require bearing plates. These two conditions are shown in Fig. 6-36.

In the better residential work, floor joists are anchored to the walls. Some cities require this by ordinance. In the great majority

135

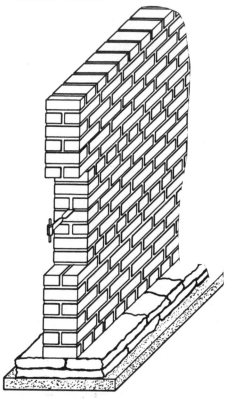

Fig. 6-33. T-anchor in an 8″ wall, which will tie into a perpendicular wall.

of residence work outside of such cities, however, anchors are not used. Anchors are spaced approximately 6 ft. apart for both floor joists and roof plate. Great care should be exercised in placing these anchors as near the bottom of the joists as possible in order to lessen the strain on the brick wall, in case a fire causes the joists to drop. Fig. 6-37 shows the right and wrong placement of joist anchors in solid walls, and Fig. 6-38 shows the correct placement in hollow walls.

In constructing the walls, the brickwork should be stopped at the point where the first floor joists are to rest on it. Care should be taken to have the top course perfectly level, so that the joists may be set without wedging or blocking. After the joists are placed, the brickwork is continued up, leaving a small "breathing" space around the joist to prevent dry rot. The same method of joisting is followed at the upper floors. On anchor joists the anchors are attached to the joists with spikes driven through the holes seen in the illustrations of anchors.

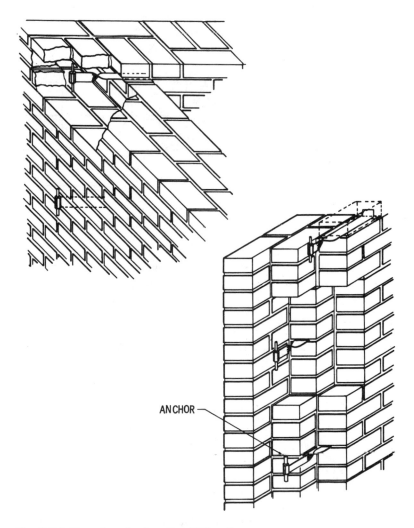

ANCHOR

Fig. 6-34. T-anchors for bonding 12″ walls.

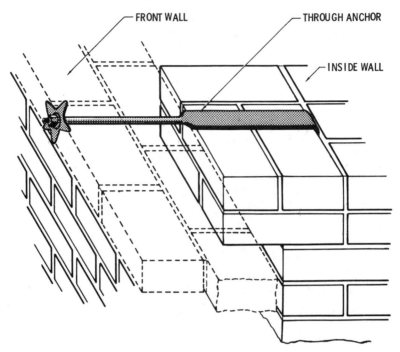

Fig. 6-35. A through anchor with nut and washer for intersecting walls.

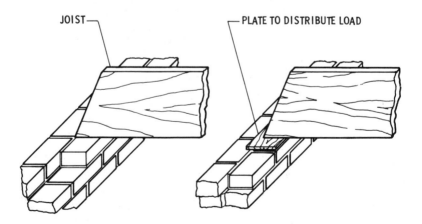

Fig. 6-36. A metal plate placed under floor joist, in some instances, to distribute weight.

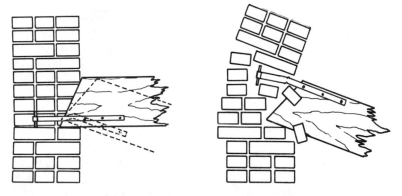

Fig. 6-37. Anchors installed to the bottom of beveled joist to prevent pulling out of bricks in case of joist breakage.

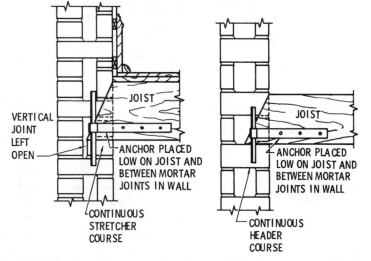

Fig. 6-38. Correct placement of anchors on joists in two types of hollow-wall construction.

The ends of all joists are beveled, whether they are anchor joists or intermediate joists, so that in case of fire they will readily fall without injury to the wall.

Anchoring the Roof Plate

Before the top of the wall is reached, the anchors for bolting down the roof plate should be placed and the brickwork carried

139

up around them as shown in Fig. 6-39. The bolts should be ½″ in diameter and at least 12″ long, with a tee or washer at the bottom and a nut and washer at the top, and should be set approximately every 6 ft. along the wall. After the carpenter has placed the roof plate and before it is bolted down, the mason should place a bed of mortar under the plate.

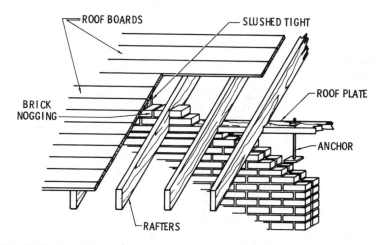

Fig. 6-39. Anchoring roof plate to brick wall.

When the wall is finally carried to the top and the roof rafters set, but before the roof boarding is in place, the mason should fill in between the roof rafters with one tier of brick, as shown. This is called *nogging.* Its purpose is to effectively block the openings between the roof rafters and to prevent the wind from entering the walls and attic. This adds greatly to the comfort of the house in cold weather. In warm climates, nogging will not be necessary.

CHAPTER 7

Corners, Openings, and Arches

As distinguished from each other, a corner is the meeting of the ends of two converging walls, whereas an intersection is the meeting of one wall with another wall at some intermediate point. In the case of intersections, one wall may end at the point of intersection or it may continue. These distinctions are shown in Fig. 7-1.

CORNERS

The corner, which is the beginning or end of a wall, is the point where the bond starts; it is here that means must be provided so that the courses may be shifted the amount required by the bond employed. This is obtained by the proper arrangement of the

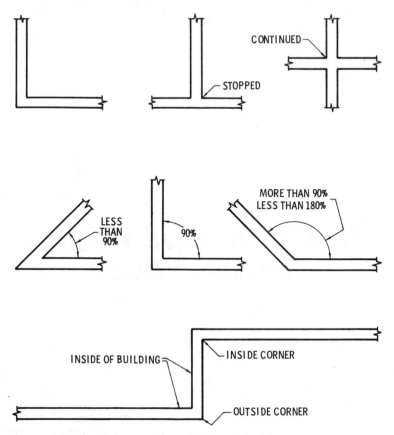

Fig. 7-1. Corners and intersections of brick walls.

brick at the corner and by the use of special brick if necessary. The various kinds of corners encountered may be classed

1. with respect to the angle of the walls, as
   a. square (90°)
   b. obtuse
   c. acute
2. with respect to the direction of angle, as
   a. outside
   b. inside

Starting the Bond at 90° Corners

To start the bond in two walls, diverging from a corner, some special arrangement of the brick is necessary; otherwise the courses would not have the required shift resulting in incorrect lap. The bricks used for this purpose are

1. quoins;
2. closers;
3. bats.

With these forms, various spacings may be obtained so as to obtain the proper lap in starting the bonds. The numerous spacing combinations that can be made are illustrated in Chapter 4. The following rule should not be violated. This rule states: *A course should be started with a quoin, never a ¼ closer*. That is, the end brick should never be less than 4″ in width. Sometimes ¼ closers are used, but this is a very bad practice and cannot be too strongly condemned.

In starting a bond at the corner, the requirements of both walls must be considered so that proper quoins and closers may be selected. The example which follows illustrates the method of solving the problem.

Example—Determine the brick arrangement at a corner for stretcher bond 4″ wall, ½ lap; ¾ lap.

Case 1, ½ lap—This is the simplest case. In Fig. 7-2, the bond starts with brick B, leaving ½ of the brick to be covered to the edge and requiring a whole brick length on the other wall. Hence, a whole quoin or end brick will fill the space.

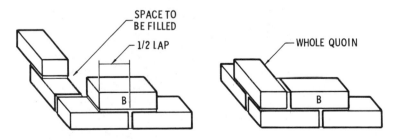

Fig. 7-2. How ½-lap corners are started. Whole bricks are used.

Case 2, ¾ lap—The first bond brick B, Fig. 7-3, will have a ¼ lap on the end stretcher M, leaving ¾ of brick M to be covered. On the other wall ¼ of brick S is to be covered, or ¾ each way. Hence, a ¾ quoin and a ¼-bat closer are required to fill the space, as shown.

The above example together with the explanations given in Chapter 4, should be sufficient to show how the brick may be arranged to meet various conditions. Both outside and inside corners for the various bonds for walls of various thickness are shown in Figs. 7-4 through 7-10.

The most frequently used wall construction is 8″. The best way to start is with a first course, over the foundation, of all header

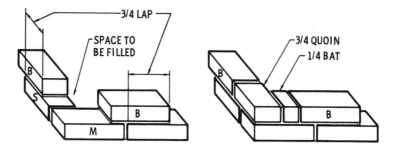

Fig. 7-3. In ¾-lap bonding, ¼-lap and ¾-quoin sizes are needed.

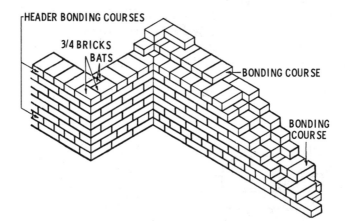

Fig. 7-4. American bond with 8″ and 12″ wall corners.

brick. Fig. 7-11 shows, step by step, how to begin the course. This is the first course and the most important in making the correct start. The dashed lines are the positions of the level. Note the frequency with which it is used. The letters show the alphabetical order in which the brick is laid.

The second step illustrated in Fig. 7-12 shows the position of the second and succeeding courses, up to about the fifth to the seventh, at which time another course of headers is placed for

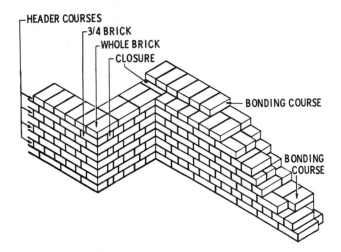

Fig. 7-5. English bond with 8″ and 12″ wall corners.

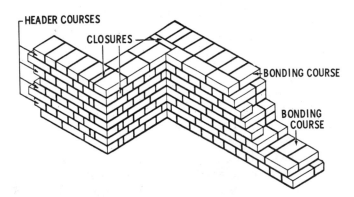

Fig. 7-6. English cross or Dutch bond corners for 8″ and 12″ walls.

145

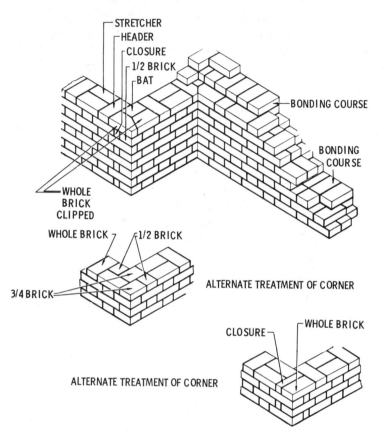

STRETCHER
HEADER
CLOSURE
1/2 BRICK
BAT
BONDING COURSE
BONDING COURSE
WHOLE BRICK CLIPPED
WHOLE BRICK
1/2 BRICK
ALTERNATE TREATMENT OF CORNER
3/4 BRICK
WHOLE BRICK
CLOSURE
ALTERNATE TREATMENT OF CORNER

Fig. 7-7. Flemish bond with 8″ and 12″ wall corners.

bonding. Fig. 7-13 shows how the outside of the corner will look with the first course all headers, five courses of stretchers, and the start of the sixth course for headers.

Fig. 7-14 is a view of the inside of the same corner. As mentioned previously, it is important that a measurement from corner to corner is accurately made and that the level is used frequently to be sure it is vertical on both sides and that all courses are horizontal. With two corners laid up, the rest of the brick is laid between them by the use of the taut line.

146

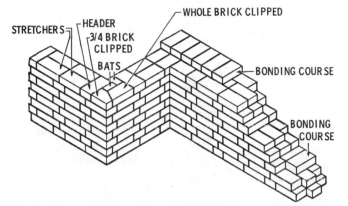

Fig. 7-8. Garden-wall corners 8″ and 12″ thick.

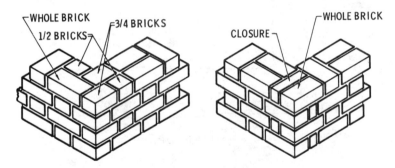

Fig. 7-9. Outside corners made with and without closers.

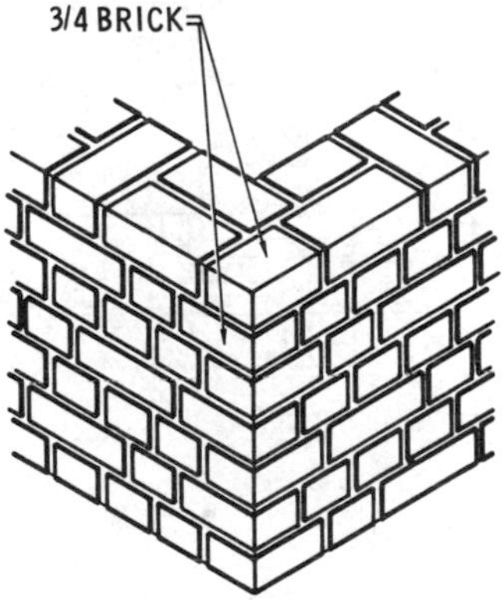

Fig. 7-10. English cross and Dutch bond corners can be started without closers.

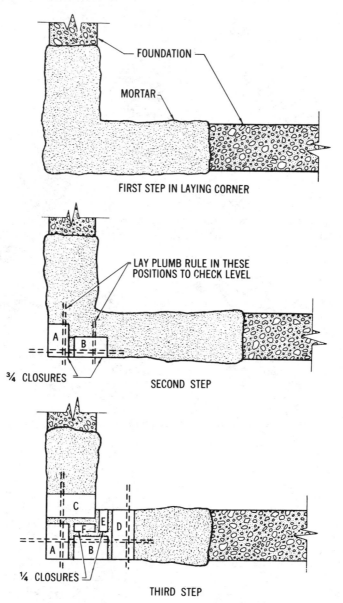

FIRST STEP IN LAYING CORNER

LAY PLUMB RULE IN THESE
POSITIONS TO CHECK LEVEL

FOUNDATION

MORTAR

A   B

¾ CLOSURES

SECOND STEP

C

E   D

F

A   B

¼ CLOSURES

THIRD STEP

Fig. 7-11. Five steps in laying

148

Obtuse-Angle Corners

If the angle of turn is 30°, 45°, or 60°, specially shaped brick made for the purpose, called *splay*, or *octagon, brick*, may be obtained from dealers and manufacturers. If for any reason these special shapes are not available, the angles may be formed by the use of standard-size brick. There are two methods of using

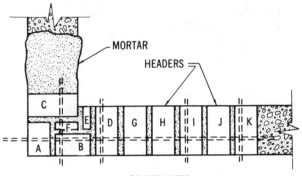

FOURTH STEP

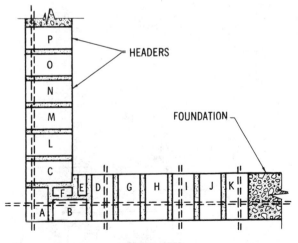

FIFTH STEP

a first course of all headers.

149

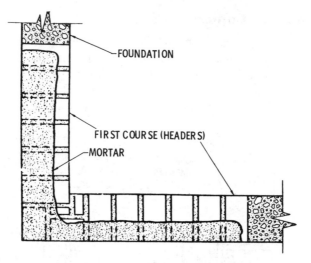

STEP 1

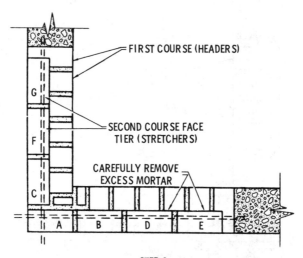

STEP 2

Fig. 7-12. Succeeding courses are common ½-lap bond up to from five to seven courses.

150

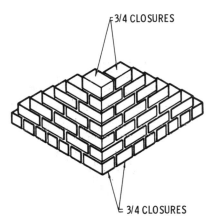

3/4 CLOSURES

3/4 CLOSURES

Fig. 7-13. The start of an American bond corner.

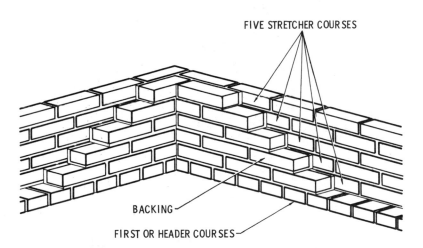

FIVE STRETCHER COURSES

BACKING

FIRST OR HEADER COURSES

Fig. 7-14. Inside tier of corner.

standard brick for outside corners, as shown in Fig. 7-15; both are objectionable, as can be seen from the illustrations.

Fig. 7-16 shows outside and inside obtuse-angle turns with standard brick in an 8″ wall. This arrangement results in a minimum amount of brick cutting. Fig. 7-17 shows a 12″ wall with minimum brick cutting. Fig. 7-18 shows shows special brick for making standard turns of 30°, 45°, and 60°.

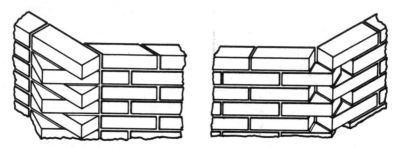

Fig. 7-15. Methods of making obtuse angles with standard brick without cutting. Neither is recommended.

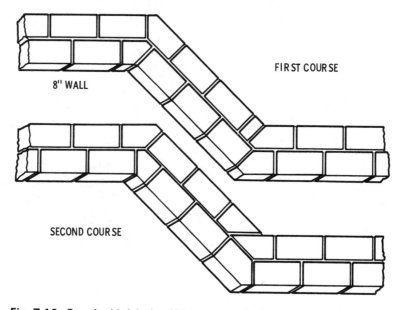

FIRST COURSE

8" WALL

SECOND COURSE

Fig. 7-16. Standard brick should be cut to make better obtuse corners.

Acute-Angle Corners

In the less expensive class of work, to save time the "pigeon-hole" method of making an acute turn is employed. In this method the wall is built out almost to a sharp edge; the brick is laid up with vacant indented spaces, or pigeonholes, on each side, as shown in Fig. 7-19.

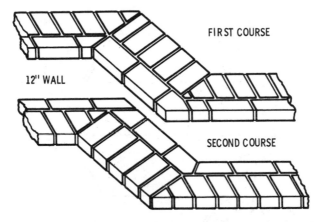

FIRST COURSE

12" WALL

SECOND COURSE

Fig. 7-17. English bond 12" wall with cut brick for obtuse corner.

The pigeonholes are the result of laying the brick with square ends instead of using special brick or chipping to shape. The pigeonholes form spaces for the accumulation of dirt, water, snow, etc., with the result that the brickwork rapidly deteriorates and, moreover, is unsightly. A better method of making an acute angle corner is shown in Fig. 7-20. Here the sharp edge and pigeonholes are avoided, thus improving the appearance and leaving no spaces for the lodging of dirt, etc.

There are two types of wall intersections:

1. Stopped or L shaped.
2. Continued or cross shaped.

These intersections are usually at right angles. The brick is arranged in various ways at these intersections, depending on the wall thickness and kind of bond. The object is to bond the two walls together while at the same time preserving the bond in each wall.

Stopped Intersections—This is the simple case of intersections, as illustrated in the intersection of a partition wall with the outer wall. In arranging the brick at the intersection, the aim should be to get the maximum amount of bonding. Fig. 7-21 shows the first and second courses of an 8" wall.

Continued Intersections—Frequently an intersecting wall continues past the intersected wall, forming a cross-shaped figure.

153

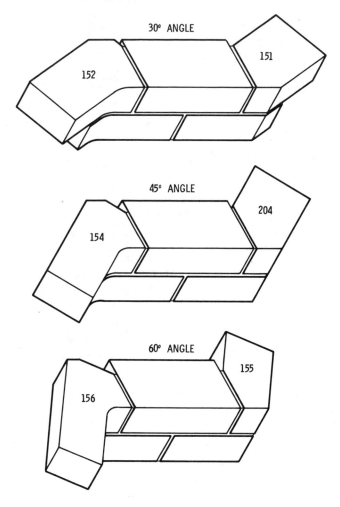

Fig. 7-18. Special pressed bricks make better obtuse corners.

The same general principles apply as for stopped intersections, the aim being:

1. To get as much bond area as possible.
2. To preserve the bond in both walls.
3. To avoid chipped brick when possible.

Fig. 7-19. "Pigeonhole" corners are developed when uncut bricks are used to construct and obtuse corner.

The structure of an 8″ English bond wall is shown in Fig. 7-22. The methods of brick arrangement for 8″ and 12″ walls, common or American bond, are shown in Figs. 7-23 and 7-24. A study of these illustrations will show the general scheme to be followed. The student should practice by applying the principles illustrated to the other bonds for walls of various thicknesses.

OPENINGS

Window and floor frames are available in standard sizes to permit placement in brick openings without the need for cutting brick to fit. This applies to the height, which allows for a whole number of courses, and to the width, to allow for a whole number of bricks for the width of the needed opening.

For years, windows were of wood sash frames and were called *double-hung* (both lower and upper windows movable). Recently, home construction has turned to aluminum and steel sash windows, with the lower window counterbalanced on springs for raising. The upper window is generally fixed. Metal sash windows include the narrow, vertically hinged windows that swing inward, allowing for the use of screens on the outside.

Door frames for homes are still made principally of wood. The large doors on office and industrial buildings are often made of metal. The bottom sill of a door is usually precast concrete. The sill of a window may be precast concrete or brick (Fig. 7-25). At the top is a lintel, usually steel, to support the weight of the brick over the openings.

In bricklaying, the problem of discontinuing the bond or stopping the brickwork for openings is not difficult when once understood. The bond should be worked out by the architect in designing the building so that the sizes and spacing of the open-

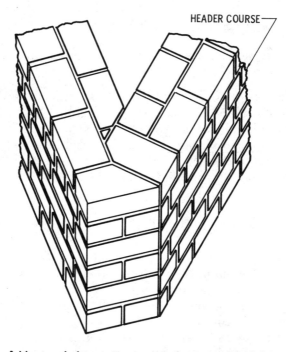

HEADER COURSE

Fig. 7-20. A blunt and obtuse corner made from standard brick.

ings will permit stopping the bond without irregular lap conditions. The brick linear dimensions, whenever possible, should be calculated so as to reduce cutting to a minimum. The competent architect will attend to this.

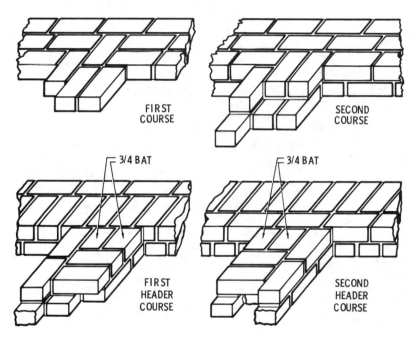

Fig. 7-21. Courses for an American bond, 12″ intersecting wall.

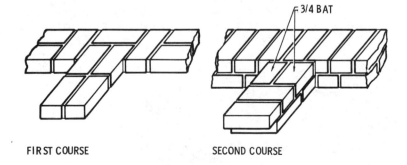

Fig. 7-22. English bond, 8″ intersecting wall.

157

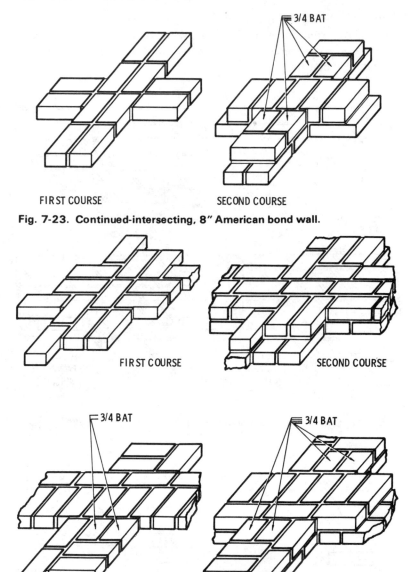

FIRST COURSE   SECOND COURSE

Fig. 7-23. Continued-intersecting, 8″ American bond wall.

FIRST COURSE   SECOND COURSE

3/4 BAT   3/4 BAT

FIRST HEADER COURSE   SECOND HEADER COURSE

Fig. 7-24. 12″ American bond, continued-intersecting wall.

158

Window Openings

Windowsills in brick buildings should be of brick or stone. Concrete, unless it is precast, is not well adapted for this purpose. Brick windowsills are preferable to stone. Brick sills add to the appearance of the building and are inexpensive since they are made of the same material as the wall and placed by the same workmen who lay up the wall, thus eliminating the necessity of additional labor to place the heavy stone. Brick for sills should be laid on edge, rowlock style, and pitched approximately at an incline of 1″ in 6″ to shed the water.

Fig. 7-25. Brick sill used at a window opening.

Fig. 7-26 shows how an opening for a window is started and the inclined brick on edge which holds the window frame. The illustration shown is for an 8″ wall. A better finish is secured by the use of special sill brick. Some of these special shaped bricks and the method of laying are shown in Fig. 7-27.

To obtain the best effect, brick sills should be "slip sills" not wider than the actual masonry opening. Brick sills laid horizontally with a pitch formed with concrete are not satisfactory since the action of the weather may cause the concrete to loosen. In

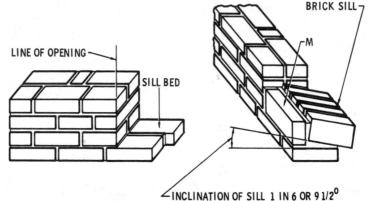

Fig. 7-26. The construction of a brick windowsill.

general, brick is the most satisfactory material for windowsills, although it may be made to form attractive combinations with other materials. Where brick is used throughout, however, no material has to be specially ordered. An appearance of great solidity may be gained by sloping the brick windowsills very sharply, thus increasing the depth of the reveal of the windows.

Windows should begin a whole number of courses above the floor level or the beginning of the brick wall. Use the 4″ module system, but remember it takes three brick courses to equal 8″, or two 4″ module dimensions. At sill height, lay the sloped rowlock course to occupy the height of two normal courses. Fig. 7-28 shows a cross section of a sill for a wood window frame and back sill. As soon as the mortar is set, put the window frame in place and brace it with boards to be sure it holds its vertical and horizontal position (Fig. 7-29). The boards should be left in place for several days. The rest of the wall is then built up to the height of the top of the window frame. If the top course does not come out at least ¼″ from the top, the difference can be adjusted with mortar.

Wood window frames must be prepainted with a prime coat before they are set into place. Stock sizes ordered from the supplier will usually be applied with a prime coat. Complete wind- and rainproofing is done at the time of installation by a thin layer of mortar. Steel frames may be plated or painted to prevent rust.

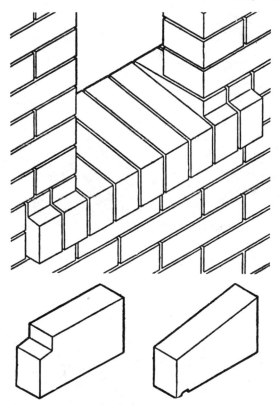

Fig. 7-27. Specially formed brick is available for windowsills.

Aluminum frames are nearly always anodized, a process that gives them a hard coat and also prevents oxidation. Anodizing can be ordered in a variety of colors.

Door Openings

The treatment of door openings is the same as for window openings, except that they begin at the bottom. A precast concrete sill is laid on the foundation, the same as the first course of brick. To ensure a firm hold onto the sides of the brick wall, pieces of wood the same size as a ½-bat brick are mortared into the side opening, the same as though they were brick. Screws are

161

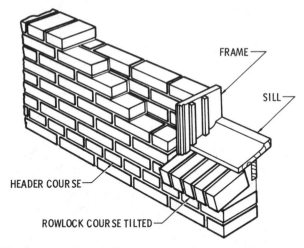

FRAME

SILL

HEADER COURSE

ROWLOCK COURSE TILTED

Fig. 7-28. Cross section of sill construction for a wood window sash.

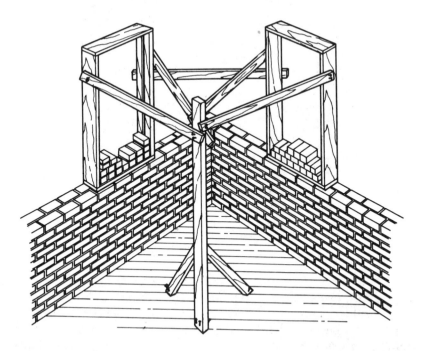

Fig. 7-29. Method of bracing wood window frames.

used later to secure the door frame. This is indicated in Fig. 7-30. Precast concrete is the most satisfactory material for door sills.

LINTELS

The support over window or other openings is formed by lintels or by arches. The treatment of arches is explained later. A lintel is the horizontal top piece of a window or doorway opening serving to support the brickwork, as in Fig. 7-31. It should be understood that the lintel is not a part of the window (or door) frame but serves to support the brickwork so that no undue stress will be brought on the frame, tending to distort it.

A wooden lintel is all the support required for brickwork over

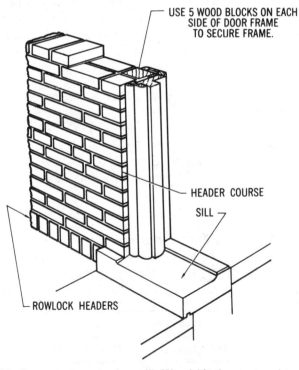

USE 5 WOOD BLOCKS ON EACH
SIDE OF DOOR FRAME
TO SECURE FRAME.

HEADER COURSE

SILL

ROWLOCK HEADERS

Fig. 7-30. Precast concrete door sill. Wood blocks mortared to the brick will later hold the door frame.

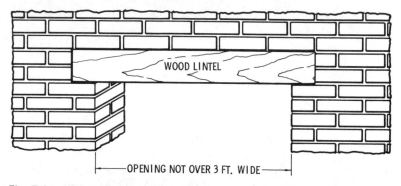

Fig. 7-31. A lintel supports the brick over an opening. Wood may be used if the opening is not over 3 ft. wide.

openings 3 ft. wide and less. When the mortar is set, brickwork will support itself over spans of this width, even though the wood lintel should burn or decay. However, since brick should never be required to support its own weight over an opening, it is preferable to use steel, precast reinforced concrete, or reinforced brick (RBM).

Fig. 7-32 is a cutaway view of a steel lintel. It consists of two angle irons, ¼″ thick and back to back. Fig. 7-33 shows the use of this type of lintel over a door opening. The ends of the lintel should extend 4″ to 8″ into the brickwork for support. The depth of the lintel into the brickwork depends on the length of the span and should be deeper for longer spans. Table 7-1 shows the lintel

Table 7-1. Lintel Sizes Recommended for Two Wall Thicknesses and Several Span Lengths

Wall thick-ness	3 feet		4 feet,* Steel angles	5 feet,* Steel angles	6 feet,* Steel angles	7 feet,* Steel angles	8 feet,* Steel angles
	Steel angles	Wood					
8″	2–3×3×¼	2×8 2–2×4	2–3×3×¼	2–3×3×¼	2–3½×3½×¼	2–3½×3½×¼	2–3½×3½×¼
12″	2–3×3×¼	2×12 2–2×6	2–3×3×¼	2–3½×3½×¼	2–3½×3½×¼	2–4×4×¼	2–4×4×4¼

*Wood lintels should not be used for spans over 3 ft. since they burn out in case of fire and allow the brick to fall.

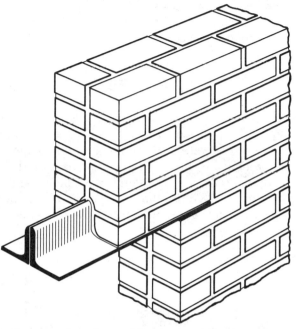

Fig. 7-32. Cutaway view of a steel lintel over an opening.

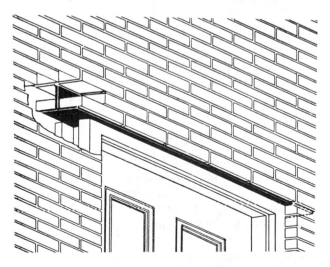

Fig. 7-33. Another type of steel lintel used over an opening.

165

sizes recommended for two wall thicknesses and several span lengths. Reinforcing rods in brickwork (Fig. 7-34) will handle moderate loads and are economical.

ARCHES

Properly constructed, an arch of brick is capable of supporting very heavy loads. Arches were used even in ancient times before the discovery of concrete or mortar as a bonding agent. Arches hewn from stone were able to support aqueducts and roads over rivers just by their shape and careful placement. While circular and elliptical (Fig. 7-35) are the most common shapes given to arches, others are jack and segmental, as shown in Fig. 7-36.

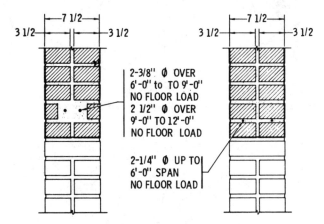

Fig. 7-34. Cutaway view of reinforcing rods to strengthen openings.

The construction of an arch should be left only to the most experienced mason. It involves a great amount of brick cutting for careful fitting of the horizontal courses to the edge of the arch and careful application of mortar to the arch bricks. Mortar should not be less than ¼″ in thickness. Because of the form of the arch, the outer edges of the brick will be farther apart and will require progressively thicker mortar joints from the inner edge to the outer edge.

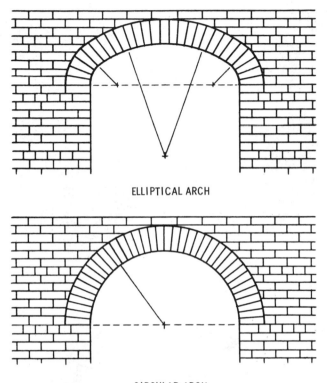

ELLIPTICAL ARCH

CIRCULAR ARCH

Fig. 7-35. The two most common arches used.

Arches of brick can be successfully placed and temporarily supported with templates or wood forms (Fig. 7-37). After carefully determining their size, two pieces of plywood are cut and fastened together with two pieces of 2″ × 4″ between. This gives two edges for the support of the brick. When the wall is built up to the beginning of the arch, the template is put into place, supported on wood posts, brick, or concrete block, as shown in Fig. 7-37. Brick is placed on edge over the arch to determine the number of bricks needed. If possible, there should be an uneven number of bricks so that the middle brick is at the center of the arch. This will be the closer brick and is the last one to be buttered with mortar and placed.

167

JACK ARCHES

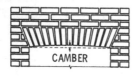
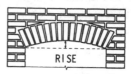

NOTE:
CAMBER 1/8" PER FOOT OF SPAN

NOTE:
MIN. RISE 1" PER FOOT OF SPAN

SEGMENTAL ARCHES

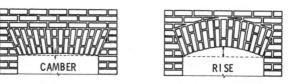

Fig. 7-36. Other popular arch contours.

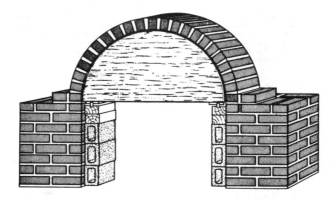

Fig. 7-37. A plywood template used to support the brick while mortar is curing.

Once the number of bricks is determined, start work from both ends, buttering and placing. In a good job the arch sections will meet in the center for placing the closer brick. Also, all mortar thicknesses will be alike. The wood form should be left in place while the rest of the courses continue up the wall around the arch.

Fig. 7-38. A double arch used in a Spanish-designed wall.

Fig. 7-39. A fireplace opening with a brick arch.

After several days, the mortar will have taken a good set and the arch will be self-supporting.

Arches continue to be favored in architectural design, especially in Spanish-style homes in the Southwest. They are used for both exterior and interior design. Fig. 7-38 shows a double arch in a Spanish-style wall. Fig. 7-39 shows a brick arch used for a fireplace opening.

CHAPTER 8

Brick Surfaces
and Patterns

Brick is not only an excellent material structurally because of its high compressive strength, but with the wide variety of patterns and colors available it has great aesthetic value. The home in Fig. 8-1 is a stucco and brick veneer home. On three sides it is stuccoed (a 1"-thick layer of concrete) over frame, which provides a sturdy and well-insulated home. The front side, however, is a single tier of brick (brick veneer) purely for its decorative value.

Fig. 8-2 is a closeup look at the brick used in this home. It carries the name "white bark." Its outer texture is like the bark of a tree, and its color is buff white rather than pure white. Standard mortar, with no color added, seems to blend well with the color of the brick.

As mentioned in Chapter 1, color is altered in brick by chemicals added and by the amount of heat and the length of time used in the kiln. Texture is the result of the processing. Figs. 8-3 to 8-5

Fig. 8-1. Brick veneer adds beauty to a home.

show a few of the variations and textures available in one small local brickyard. They may or may not be available in other brickyards. When planning a new home or building, one should review the brick colors and textures that can be obtained locally.

Fig. 8-2. Appearance of the texture in the brick used in the home in Fig. 8-1.

172

Fig. 8-3. Common brick with different amounts of oxide added to produce different colors.

Fig. 8-4. Deep vertical scoring in brick can alter apparent dimensions of a wall.

A large brick processor will produce a large variety of bricks. If the processor is located at some distance from your city, expect to pay more for the brick because of the high cost of transportation of such heavy material.

Brick probably offers the greatest flexibility in decorative variations, at a reasonable cost, of any building material of comparable strength. Because brick is available in such a variety of colors, it lends itself easily to the formation of patterns with endless variations.

HOW TO MAKE PATTERNED WALLS

Chapter 4 described the standard patterns resulting from bonding in a number of traditional forms. Included were garden walls (Fig. 8-6) representing variations of the standard patterns. With imagination and artistic talent, an almost infinite variety of patterns becomes possible without loss of good bonding.

A great variety of ornamental figures may be obtained by the use of light and dark bricks or bricks of various shades, and by

Fig. 8-5. Brick can be made to look like natural stone.

Fig. 8-6. Diagonal lines and dovetail patterns form a garden-wall pattern with the help of brick color variations.

the combination of headers and stretchers in each course variously arranged.

Stretcher Bond Patterns

With the alternate light and dark bricks, various effects may be produced, depending on:

1. Lap.
2. Shift.
3. Relative position of shades in adjacent course, as shown in Fig. 8-7.

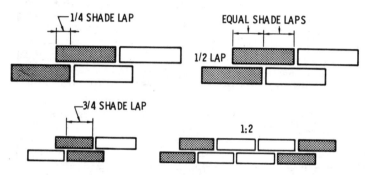

Fig. 8-7. Various diagonal patterns may be formed by altering the lap.

A variation consists of laying two light bricks to one dark brick (Fig. 8-8), thus further separating the dark diagonals. Various other combinations of the two shades may be made, securing additional effects.

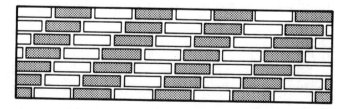

Fig. 8-8. A diagonal pattern in a wall using ¼ lap.

Header Bond Patterns

Similar two-shade effects may be obtained in header courses, but only with ¼ lap. The general appearance of a wall laid with two-shade headers alternating is indicated in Fig. 8-9.

Fig. 8-9. Appearance of a wall laid with header bond, two shade, ¼ lap.

Reversal of Shift

The monotony of long "dark diagonals" may be broken by the simple device of reversing the shift, as indicated in Fig. 8-10. The appearance of a wall laid in header bond with reversal of shift is shown in Fig. 8-11.

Various patterns may be obtained by the use of dark brick for:

1. Headers.
2. Stretchers.
3. Both headers and stretchers in combination with various shade shifts, reversal of shifts, etc.

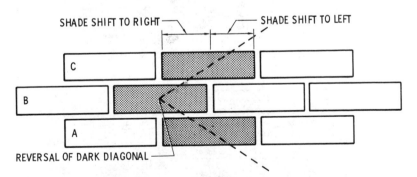

Fig. 8-10. Reversing the line shift can break up the monotony of a continuous diagonal line.

Fig. 8-11. Appearance of a wall with reverse shift of shading.

Effects may be obtained in this combination of stretcher and header bonds in various ways. Thus, dark horizontals at uniform intervals are obtained with dark headers; also V's and even diamonds are made by proper placing of dark bricks.

Many patterns are obtainable in Flemish bond (single and double) using the bonds just described, dark headers, dark stretchers, or combination of both in groups. The Flemish bond with dark headers gives dark uprights and dark diagonals combined, as shown in Fig. 8-12, which also illustrates the single and double form of bonding.

Garden-Wall Bond Patterns

Owing to the increased number of stretchers used in this bond, the patterns obtained are more extended than in the other bonds. Fig. 8-13 shows appearance of the three-stretcher form with dark

Fig. 8-12. Flemish bond with shaded headers.

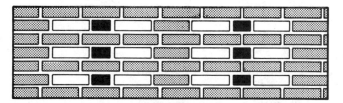

Fig. 8-13. A three-stretcher garden wall with dark headers.

headers, the latter centered over stretchers. A three-shade effect with garden-wall courses alternating with stretcher courses is shown in Fig. 8-14.

Diamond Unit Patterns

Fig. 8-15 shows the unit system. The diamond shapes formed by various combinations of headers and stretchers represent the various units, or "eyes," upon which all diagonal bonds are based.

Beginning with unit 1, which is composed of a stretcher with a header centered above and below it, each succeeding unit is formed by extending every course of the preceding unit the width of a header, always centering the courses on the middle course, regarded as the horizontal axis of the unit, and terminating the whole above and below by a header. As a result, no matter how far they may be carried out, they always present exact mathematical proportions and bear a definite relationship to each other.

Fig. 8-14. A garden wall with bricks of three different shades.

It is interesting to note that the units may also be recognized by their horizontal axes, which in odd-numbered units are always composed entirely of stretchers, while in even-numbered units always carry one, and only one, header set as near their center as

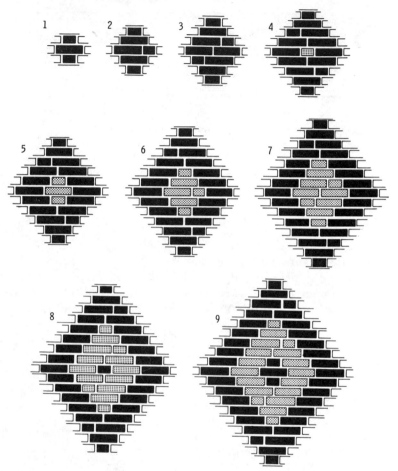

Fig. 8-15. Diamond pattern of various sizes may be developed by a unit system.

possible. The number of an even-numbered unit is double the number of stretchers in its axis, while that of an odd-numbered unit is one less than double the number of its axial stretchers. Thus, if we see a unit with a horizontal axis of 4 stretchers only, we may be sure that it is the odd-numbered unit 7; but if it has 4 stretchers and a header, we know that it is the even-numbered unit 8.

180

With unit 4 (Fig. 8-15), there begin to appear units within units. While the header, crossed by vertical stretcher joints which appear at the center of unit 4, is not strictly a unit in our sense of the term, it is nevertheless the primary unit of all as the smallest normal element in brickwork. Unit 1 clearly comes to view as the center of unit 5; unit 2 appears in 6; unit 3 in 7; and so on. It is by the treatment of these units, each of which in itself is a bond pattern, that various patterns may be worked out on the surface of the wall by the proper handling of the shades and textures in the brick or of the mortar joints.

The units may be made to join, or butt, each other vertically and horizontally; or they may be separated by introducing between them one or more courses above and below, or variously arranged rows of brick in a general vertical direction on the side, as may be seen in Fig. 8-15. When separated, the units are said to be surrounded by horizontal and vertical borders. Much of artistic value of the pattern will depend on the skill and taste with which these borders are worked out.

The brickwork designer is urged to remember that the use of pattern bonds requires the strictest and most thoughtful attention to the beginning and ending of the pattern at either the bottom or top of the structure or on piers as they occur separately or between windows. He must first decide on a unit which is suitable to the size of the panel to be covered and then exactly center it on the panel so that his pattern may end in a symmetrical manner, both laterally and vertically. In order to secure vertical symmetry, the panel must always have an odd number of courses, that is, an even number on each side of the median line.

Treatment of the Units

Fig. 8-15 shows the appearance of the units in a wall pattern and indicates the way to treat the shades and textures of brick in designing patterns. The illustrations are by no means intended to dictate what may or may not be the best shade combinations in any given bond, or the blendings of light and shade, or the contrasts of color and texture in brickwork; they merely suggest pattern designs.

At the same time, an underlying principle is involved in what has been pictured. It will be readily understood that the smaller units which are worked in to patterns of finer texture and quieter shadings are most appropriate for walls of limited area, while the bolder outlines and heightened contrasts of the larger figures are more suitable for larger sweeping wall surfaces.

Brick Pattern Diagrams

To assist the bricklayer in laying out pattern designs, a diagram or drawing of the brick units in the wall surface may be made. The brick manufacturers usually use especially designed cross-section paper, which is ruled with spaces equal in length to a stretcher and a mortar joint.

This sectional paper is laid out in such a way that the actual dimensions have been reduced to $\frac{1}{16}$ of the original. The length is also divided into four equal parts so that the overlapping of the brick into the various courses may be properly located. The course heights are laid out by using the depth or thickness of a brick and the width of a mortar joint.

Fig. 8-16 shows a drawing on which the stretchers and course heights of brick are indicated by definite units of measurement. A diagonal pattern, designed in the upper part of the diagram, is laid out from a center line so that the bond pattern is symmetrical on each side of the center.

In addition to laying out the pattern bond on paper, the brick-layer should compute the length of courses in feet and inches and the total height of various pattern bonds with varying widths of mortar joints in feet and inches.

This distance can be determined in a wall laid up in running bond by multiplying the unit length of a brick and mortar joint by the number of bricks in the course. In other types of bonds, such as the Flemish, where the courses are made up of stretcher and header, the problem is not quite so simple.

Tables, scales, and bricklayers' rules, however, have been devised to assist the bricklayer in determining the number of bricks in a course of a certain length and the number of courses in a definite height.

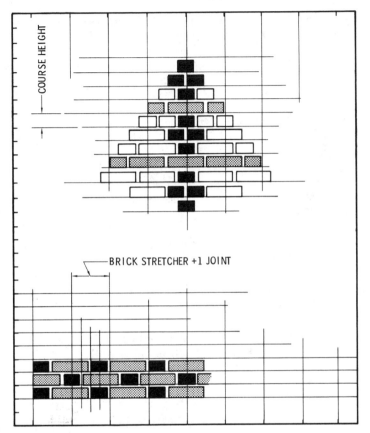

Fig. 8-16. Cross-section paper is an aid to developing patterns before the start of actual brickwork.

MORTAR JOINTS

In addition to variations in brick, mortar joints may also be treated in a number of ways to produce different artistic effects. Mortar-joint finishes fall into two classes: troweled and tooled joints. In the troweled joint, the excess mortar is simply cut off (struck) with a trowel and finished with the trowel. For the tooled joint, a special tool other than the trowel is used to com-

183

press and shape the mortar in the joint. Fig. 8-17 shows a cross section of typical mortar joints used in good brickwork.

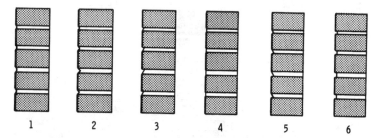

Fig. 8-17. Mortar joint treatment must be considered in brick masonry appearance.

Concave Joint (1) and V-Shaped Joint (2)—These joints are normally kept quite small and are formed by the use of a steel jointing tool. These joints are very effective in resisting rain penetration and are recommended for use in areas subjected to heavy rains and high winds.

Weathered Joint (3)—This joint requires care as it must be worked from below. However, it is the best of the troweled joints because it is compacted and sheds water readily.

Rough-Cut or Flush Joint (4)—This is the simplest joint for the mason since it is made by holding the edge of the trowel flat against the brick and cutting in any direction. This produces an uncompacted joint with a small hairline crack where the mortar is pulled away from the brick by the cutting action. This joint is not always watertight.

Struck Joint (5)—This is a common joint in ordinary brickwork. Since the mason usually works from the inside of the wall, this is an easy joint to strike with a trowel. Some compaction occurs, but the small ledge does not shed water readily, resulting in a less permeable joint than some others.

Raked Joint (6)—This is made by removing the surface of the mortar while it is still soft with a square-edged tool. While the joint may be compacted, it is difficult to make weathertight and is not recommended where rains, high winds, or freezing are likely to occur. This joint produces marked shadows and tends to darken the overall appearance of the wall.

Colored mortars may be successfully used to enhance the patterns in masonry. Two methods are commonly used: First, the entire mortar joint may be colored; second, where a tooled joint is used, tuck pointing is the best method. In this technique, the entire wall is completed with a 1"-deep raked joint, and the colored mortar is carefully filled in later.

CHAPTER 9

Chimneys
and Fireplaces

The term *chimney* generally includes both the chimney proper and (in house construction) the fireplace. There is no part of a house that is more likely to be a source of trouble than a chimney that is improperly constructed. Accordingly, it should be built so that it will be mechanically strong and properly shaped and proportioned to give adequate draught.

For strength, chimneys should be built of solid brickwork and should have no openings except those required for the heating apparatus. If a chimney fire occurs, considerable heat may be generated in the chimney, and the safety of the house will then depend on the integrity of the flue wall. Care in the construction of fireplaces and chimneys will prove to be the best insurance. As a first precaution, all wood framing of floors and the roof must be kept at least 2″ away from the chimney, and no other woodwork of any kind should be projected into the brickwork surrounding the flues (Fig. 9-1).

When it is understood that the only power available to produce a natural draught in a chimney is due to the small difference in weight of the column of hot gases in the chimney and of a similar column of cold air outside, the necessity of properly constructing the chimney so that the flow of gases will encounter the least resistance is emphasized.

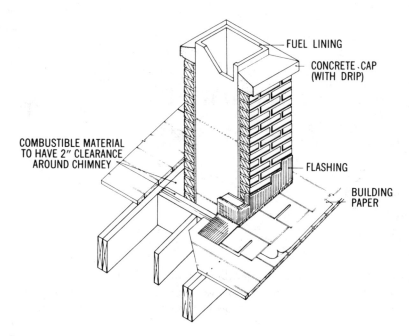

FUEL LINING

CONCRETE CAP (WITH DRIP)

COMBUSTIBLE MATERIAL TO HAVE 2″ CLEARANCE AROUND CHIMNEY

FLASHING

BUILDING PAPER

Fig. 9-1. Chimney construction above the roof.

The intensity of chimney draught is measured in inches of a water column sustained by the pressure produced and depends on:

1. The difference in temperature inside and outside the chimney.
2. The height of the chimney.

Theoretical draft in inches of water at sea level is as follows: Let,

D = theoretical draft
H = distance from top of chimney to grates
T = temperature of air outside of chimney
T_1 = temperature of gases in the chimney.

then

$$D = 7.00H \left(\frac{1}{461 + T} - \frac{1}{461 + T_1} \right)$$

The results obtained represent the theoretical draft at sea level.

For higher altitudes, the calculations are subject to correction as follows:

For altitudes (in feet) of	Multiply by
1,000	0.966
2,000	0.932
3,000	0.900
5,000	0.840
10,000	0.694

A frequent cause of poor draught in house chimneys is that the peak of the roof extends higher than the chimney. In such cases the wind sweeping across or against the roof will form eddy currents that drive down the chimney or check the natural rise of the gases, as shown in Fig. 9-2. To avoid this, the chimney should be extended at least 2 ft. higher than the roof, as shown in Fig. 9-3.

In order to reduce to a minimum the resistance or friction due to the chimney walls, the chimney should run as near straight as possible from bottom to top. This not only gives better draught but facilitates cleaning. If, however, offsets are necessary from one story to another, they should be very gradual. The offset should never be displaced so much that the center of gravity of the upper portion falls outside the area of the lower portion. In other words, the center of gravity must fall within the width and thickness of the chimney below the offset.

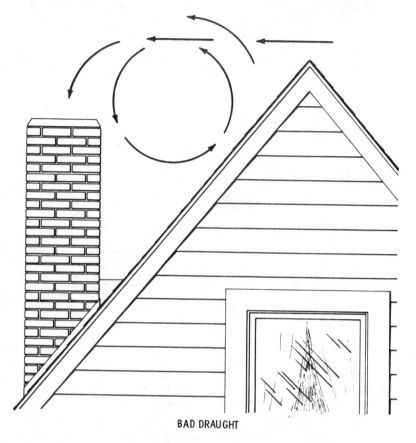

BAD DRAUGHT

Fig. 9-2. How a roof peak higher than the top of the chimney can cause down drafts.

FLUES

A chimney serving two or more floors should have a separate flue for every fireplace. The flues should always be lined with some fireproof material. In fact, the building laws of large cities provide for this. The least expensive way to build these flues is to make the walls 4″ thick lined with burned clay flue lining. With walls of this thickness, never omit the lining and never replace the

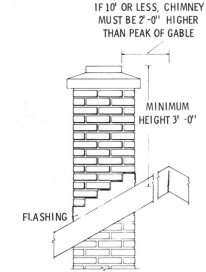

IF 10' OR LESS, CHIMNEY
MUST BE 2'-0" HIGHER
THAN PEAK OF GABLE

MINIMUM
HEIGHT 3'-0"

FLASHING

Fig. 9-3. Ample clearance is needed between peak of roof and top of chimney.

lining with plaster. The expansion and contraction of the chimney would cause the plaster to crack and an opening from the interior of the flue would be formed. See that all joints are completely filled with mortar.

Flue Lining

Walls 8″ or more in thickness may be used without a flue lining. However, walls under 8″ must have a lining of fired clay. With the increased use of gas furnaces in homes, fired-clay linings are recommended for all chimneys because of chemical action of gas residue on common brick.

Clay lining for flues also follows the modular system of sizes. Table 9-1 lists common sizes currently available. The flue lining should extend the entire height of the chimney, projecting about 4″ above the cap and a slope formed of cement to within 2″ of the top of the lining, as shown in Fig. 9-3. This helps to give an upward direction to the wind currents at the top of the flue and tends to prevent rain and snow from being blown down inside the chimney.

191

Table 9-1. Standard sizes of Modular Clay Flue Linings

Minimum Net Inside Area (sq. in.)	Nominal* Dimensions (in.)	Outside† Dimensions (in.)	Minimum Wall Thickness (in.)	Approximate Maximum Ouside Corner Radius (in.)
15	4 × 8	3.5 × 7.5	0.5	1
20	4 × 12	3.5 × 11.5	0.625	1
27	4 × 16	3.5 × 15.5	0.75	1
35	8 × 8	7.5 × 7.5	0.625	2
57	8 × 12	7.5 × 11.5	0.75	2
74	8 × 16	7.5 × 15.5	0.875	2
87	12 × 12	11.5 × 11.5	0.875	3
120	12 × 16	11.5 × 15.5	1.0	3
162	16 × 16	15.5 × 15.5	1.125	4
208	16 × 20	15.5 × 19.5	1.25	4
262	20 × 20	19.5 × 19.5	1.375	5
320	20 × 24	19.5 × 23.5	1.5	5
385	24 × 24	23.5 × 23.5	1.625	6

*Cross section of flue lining shall fit within rectangle of dimension corresponding to nominal size.
†Length in each case shall be 24 ± 0.5 in.

The information given here is intended primarily for chimneys on residential homes. They will usually carry temperatures under 600°F. Larger chimneys, used for schools and other larger buildings, have a temperature range between 600 and 800°F. Industrial chimneys with temperatures above 800°F often are very high and require special engineering for their planning and execution. High-temperature brick chimneys must include steel reinforcing rods to prevent cracking due to expansion and contraction from the changes in temperature.

CHIMNEY CONSTRUCTION

Every possible thought must be given to providing good draft, leakproof mortaring, and protection from heat transfer to combustible material. Good draft means a chimney flue without obstructions. The flue must be straight from the source to the outlet. Metal pipes from the furnace into the flue must end flush with the inside of the chimney and not protrude into the flue, as shown in Fig. 9-4. The flue must be straight from the source to

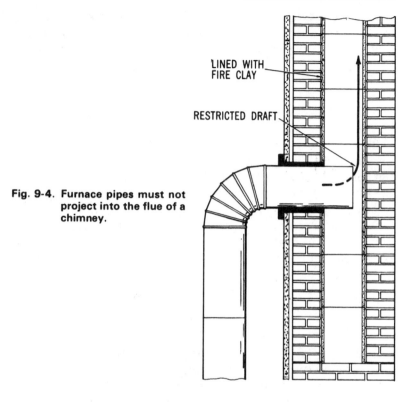

LINED WITH
FIRE CLAY

RESTRICTED DRAFT

Fig. 9-4. Furnace pipes must not project into the flue of a chimney.

the outlet, without any bends, if at all possible. When two sources, such as a furnace and a fireplace, feed the one chimney, they each must have separate flues.

To prevent leakage of smoke and gas fumes from the chimney into the house and to improve the draft, a special job of careful mortaring must be done. The layer of mortar on each course of brick must be even and completely cover the bricks. End buttering must be complete. However, it is best to mortar the flue lining lightly between the lining and the brick. Use just enough to hold the lining securely. The air space that is left acts as additional air insulation between the lining and the brick and reduces the transfer of heat.

No combustible material, such as the wood or roof rafters or floor joists, must abut the chimney itself. There should be at least a 2″ space between the wood and the brick of the chimney, as shown in Fig. 9-1. Bricks and flue lining are built up together. The lining clay is placed first, and the bricks built up around it. Another section of lining is placed, and brick built up, etc.

Chimneys carrying away the exhaust of oil- and coal-burning furnaces, where still used, need a cleanout trap. An air-tight, cast-iron door is installed at a point below the entrance of the furnace smoke pipe.

Because of the heavier weight of the brick in a chimney, the base must be built to carry the load. A foundation for a residential chimney should be about 4″ thick. If a fireplace is included, the foundation thickness should be increased to 8″.

After the chimney has been completed, it should be tested for leaks. Build a smudge fire in the bottom, and wait for smoke to come out of the top. Cover the top, and carefully inspect the rest of the chimney for leaks. If there are any, add mortar at the points of leakage.

FIREPLACES

As a way of saving energy, a fireplace is definitely not the way to go—the fire uses house air to burn, and if that air is warmed by the furnace, you really *waste* fuel. But there is nothing like the flickering flames of a wood-burning fireplace to bring cheer to the hearts of those sitting around it. Most new construction for single-family homes includes a fireplace when the house is built. A fireplace can be added to just about any home that is not so equipped.

There was a time when the fireplace or an open fire was the sole source for cooking and heating. In a few areas of the United States where the winter climates are too mild for a heating system, the fireplace is still the only source of heat to take the chill off a cool evening.

The fireplace dates back to the earliest history of man. The first home fires, forerunners to the modern fireplace, were those kindled on the earth or on a conveniently placed slab stone around which the family gathered to prepare its food. In just what period in our history fires were first used will perhaps never be known. We have evidence that primitive man made use of caves in his first temporary dwelling and built fires at the mouth of these caves, not only to prepare food, but also to protect his family from enemies.

Later, when dwellings were constructed outside of caves, family life centered in one large room in the middle of which a wood fire was lighted. Here the smoke was allowed to escape as best it could through a hole in the roof or crevices in the wall. This use of fire for heating and cooking was adopted even by the nomads, who built fires in the center of their tents and allowed the smoke to escape through a prepared opening at the top.

As more permanent and larger habitations were built and balconies or second floors were used for sleeping quarters, the hearthstone was moved to the corner of the room and an opening made in the wall to allow the smoke to escape. Later a stone hood which sloped back against the wall was added to aid in carrying the smoke out of the building.

Gradually the efficiency of the open fire was increased, and eventually the fireplace was constructed in a recess in the center of one the walls, with its own hood and enclosed flue leading up to a chimney on top of the wall. As time passed, more consideration was given to the comforts of living. The fireplace was not only improved, but became the central decorative feature of the home.

The value of fireplaces was appreciated in England as early as the latter part of the fourteenth century, when they became ornamental features in the better homes. Count Rumford, an English scientist who published a series of essays on chimneys and fireplaces in 1796, is the one to whom we are most indebted for the improvement in fireplace design and for the rules governing the openings and flues. He spent a great deal of time studying the errors of fireplace construction and the principles governing the circulation of gases and combustion.

Probably in no other country have so many types and styles of fireplaces been constructed as in the United States. Although the ornamental mantel facings of fireplaces may be of other materials than brick, the chimney and its foundation are usually of masonry construction. Fig 9-5 shows a cross section of a fireplace and chimney stack suitable for the average home.

195

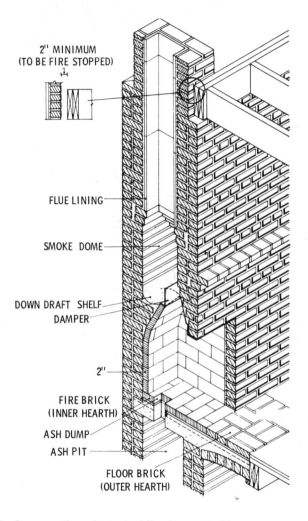

2" MINIMUM
(TO BE FIRE STOPPED)

FLUE LINING

SMOKE DOME

DOWN DRAFT SHELF
DAMPER

2"

FIRE BRICK
(INNER HEARTH)

ASH DUMP

ASH PIT

FLOOR BRICK
(OUTER HEARTH)

Fig. 9-5. Cross section of a typical fireplace and chimney.

Fireplaces are generally built in the living room or den. Newer homes have a den, in addition to the living room, and they are frequently wood-paneled. The location of the fireplace should allow the maximum of heat to be radiated into the room, with consideration given to making it the center of a conversation area. At one time its location was dictated by the location of the furnace to make use of a common chimney. In modern construction, with heating systems of the compact gas-fired type, the chimney is often a metal pipe from the furnace flue, straight through to the roof, not of the brick construction type. The brick fireplace chimney can then be placed to suit the best fireplace location in the home and room.

Fireplace styles vary considerably, from a rather large one with a wide opening (Fig. 9-6) to a smaller corner fireplace, such as the Spanish style shown in Fig. 9-7. You can also get a variety,

Fig. 9-6. A fireplace is cheery and warming but a big energy waster.

197

Fig. 9-7. Spanish-style fireplace.

detailed later, which are prefabricated and lend themselves to do-it-yourself installation.

Although large pieces of wood can be burned in the larger fireplaces, regardless of size, experience has indicated certain ratios of height, width, depth, etc., should be maintained for best flow of air under and around the burning wood. Recommended dimensions are shown in Table 9-2, which are related to the sketches in Fig. 9-8.

FIREPLACE CONSTRUCTION

Brick masonry is nearly always used for fireplace construction. Sometimes brick masonry is used around a metal form frequently given the name *heatilator*. While any type of brick may be used for the outside of the fireplace, the fire pit must be lined with a high-temperature fire clay or fire brick.

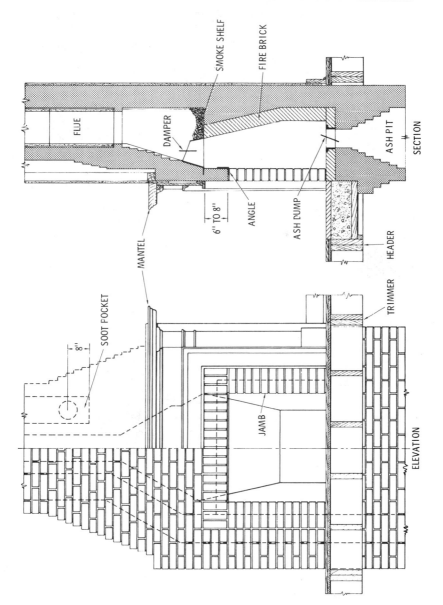

Fig. 9-8. Sketch of a basic fireplace. (Letters refer to sizes recommended in Table 9-2.)

199

Table 9-2. Recommended Sizes of Fireplace Openings

Opening			Mini-mum back (hori-zontal) c	Vertical back wall, a	Inclined back wall, b	Outside di-mensions of standard rectangular flue lining	Inside diameter of standard round flue lining
Width, w (in.)	Height, h (in.)	Depth d (in.)	(in.)	(in.)	(in.)	(in.)	(in.)
24	24	16—18	14	14	16	8½ by 8½	10
28	24	16—18	14	14	16	8½ by 8½	10
24	28	16—18	14	14	20	8½ by 8½	10
30	28	16—18	16	14	20	8½ by 13	10
36	28	16—18	22	14	20	8½ by 13	12
42	28	16—18	28	14	20	8½ by 18	12
36	32	18—20	20	14	24	8½ by 18	12
42	32	18—20	26	14	24	13 by 13	12
48	32	18—20	32	14	24	13 by 13	15
42	36	18—20	26	14	28	13 by 13	15
48	36	18—20	32	14	28	13 by 18	15
54	36	18—20	38	14	28	13 by 18	15
60	36	18—20	44	14	28	13 by 18	15
42	40	20—22	24	17	29	13 by 13	15
48	40	20—22	30	17	29	13 by 18	15
54	40	20—22	36	17	29	13 by 18	15
60	40	20—22	42	17	29	18 by 18	18
66	40	20—22	48	17	29	18 by 18	18
72	40	22—28	51	17	29	18 by 18	18

The pit is nearly always sloped back and generally sloped on the sides. This is to reflect forward as much of the heat as possible. The more surface exposure that is given to the hot gases given off by the fire, the more heat will be radiated into the room. Fig. 9-5 is a cutaway view of an all-brick fireplace for a home with a basement. The only nonbrick item is the adjustable damper. A basement makes possible a very large ash storage before cleanout is necessary. The ash dump opens into the basement cavity. A cleanout door at the bottom opens inward into the basement.

Fig. 9-9 is a side cutaway view of a typical fireplace for a home built on a concrete slab. It uses the metal form mentioned. The ashpit is a small metal box which can be lifted out, as shown in Fig. 9-10. In some slab home construction, the ashpit is a cavity formed in the concrete foundation with an opening for cleanout

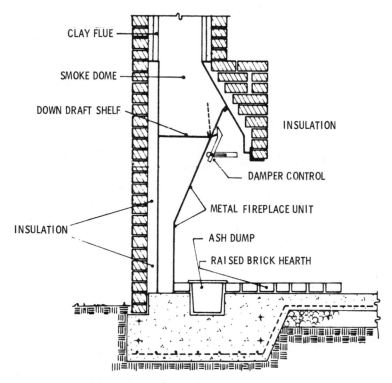

CLAY FLUE

SMOKE DOME

DOWN DRAFT SHELF

INSULATION

DAMPER CONTROL

METAL FIREPLACE UNIT

INSULATION

ASH DUMP

RAISED BRICK HEARTH

Fig. 9-9. Fireplace built on a concrete slab.

at the rear of the house. A metal grate over the opening prevents large pieces of wood from dropping into the ashpit, as shown in Fig. 9-11.

Importance of a Hearth

Every fireplace should include a brick area in front of it where hot wood embers may fall with safety. The plan view of Fig. 9-12 shows a brick hearth built 16″ out from the fireplace itself. This should be about the minimum distance. Most often the hearth is raised several inches above the floor level. This raises the fireplace itself, all of which makes for easier tending of the fire.

Fig. 9-10. Metal liftout ash box used in many fireplaces built on a concrete slab.

Fig. 9-11. A cast-iron grate over the ash box to keep large pieces of burning wood from falling into the ash box.

202

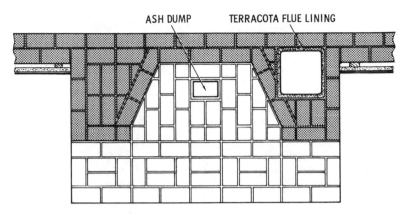

ASH DUMP   TERRACOTA FLUE LINING

Fig. 9-12. A brick hearth in front of the fireplace catches hot embers that may fall out of the fire.

In addition to the protection of the floor by means of a hearth, every wood-burning fireplace should have a screen to prevent flying sparks from being thrown beyond the hearth distance and onto a carpeted or plastic tile floor.

Ready-Built Fireplace Forms

There are a number of metal forms available, which make fireplace construction much easier. They make an ideal starting point for the handy homeowner who can build his own fireplace addition to his home. They may be called by several names—heatilator or modified forms (Fig. 9-13).

These units are built of heavy metal or boiler plate steel and designed to be set into place and concealed by the usual brickwork, or other construction, so that no practical change in the fireplace mantel design is required by their use. One advantage claimed for modified fireplace units is that the correctly designed and proportioned fire box manufactured with throat, damper, smoke shelf, and chamber provides a form for the masonry, thus reducing the risk of failure and ensuring a smokeless fireplace.

There is, however, no excuse for using incorrect proportions; and the desirability of using a foolproof form, as provided by the modified unit, is not necessary merely to obtain good propor-

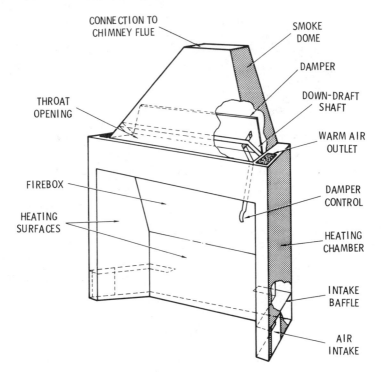

CONNECTION TO CHIMNEY FLUE

SMOKE DOME

DAMPER

THROAT OPENING

DOWN-DRAFT SHAFT

WARM AIR OUTLET

FIREBOX

DAMPER CONTROL

HEATING SURFACES

HEATING CHAMBER

INTAKE BAFFLE

AIR INTAKE

Fig. 9-13. A prefabricated metal form which makes fireplace construction easier.

tions. Each fireplace should be designed to suit individual requirements; and if correct dimensions are adhered to, a satisfactory fireplace will be obtained.

Prior to selecting and erecting a fireplace, several suitable designs should be considered and a careful estimate of the cost should be made. It should also be borne in mind that even though the unit of a modified fireplace is well designed, it will not operate properly if the chimney is inadequate. Therefore, for satisfactory operation, the chimney must be made in accordance with the rules for correct construction to give satisfactory operation with the modified unit as well as with the ordinary fireplace.

Manufacturers of modified units also claim that labor and materials saved tend to offset the purchase price of the unit, and that the saving in fuel tends to offset the increase in first cost. A minimum life of 20 years is usually claimed for the type and thickness of metal commonly used in these units.

Figs. 9-14 to 9-17 show how the brickwork is built up around a metal fireplace form. Note the ash door in Fig. 9-14, which gives access to the ashpit for removal from the outside of the house.

Fig. 9-15 shows a partially built front view. By leaving a large air cavity on each side of the metal form and constructing the brickwork with vents, some of the heat passing through the metal sides will be returned to the room. The rowlock stacked brick with no mortar, but an air space, permits cool air to enter below and warmed air to come out into the room from the upper outlet.

The front of the form includes a lintel for holding the course of brick just over the opening. A built-in damper is part of the form. Even with the use of a form, a good foundation is necessary for proper support as there is still quite a bit of brick weight. Chimney construction, following the illustrations and descriptions previously given, is still necessary.

Other Fireplace Styles

There are a number of other fireplace styles available. One is the hooded type, which permits the construction of the fireplace out into the room rather than into the wall (Fig. 9-18).

Prefabricated fireplaces mentioned earlier, come in a tremendous variety of sizes and shapes. They are usually metal and go together like an erector set. All the sizing is done. You can get them free-standing for installation against a wall (with appropriate fire safeguards) and shaped to fit in a corner. Finishing off hearts with brick or the like can lend them a built-from-scratch look.

Another is a two-sided fireplace similar to that shown in Fig. 9-19. It is used for building into a semidivider-type wall, such as between a living area and a dining area. Thus, the fire may be enjoyed from either room, or both at once, and what heat is given off is divided between the two areas.

Important to successful wood burning is good circulation of air under and around the sides. A heavy metal grate which lifts the

Fig. 9-14. Brickwork around the back of a fireplace form.

Fig. 9-15. Front view of the brickwork around a metal fireplace form.

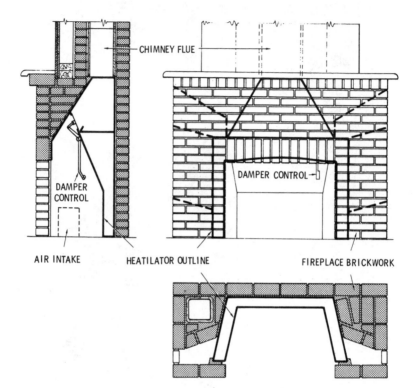

CHIMNEY FLUE

DAMPER CONTROL

DAMPER CONTROL

AIR INTAKE   HEATILATOR OUTLINE   FIREPLACE BRICKWORK

Fig. 9-16. Sketch of a typical fireplace built around a metal form.

burning logs above the floor of the fireplace is essential (Fig. 9-20).

SMOKY FIREPLACES

When a fireplace smokes, it should be examined to make certain that the essential requirements of construction, as previously outlined, have been fulfilled. If the chimney has not been stopped up with fallen brick and the mortar joints are in good condition, a survey should be made to ascertain that nearby trees or tall buildings do not cause eddy currents down the flue.

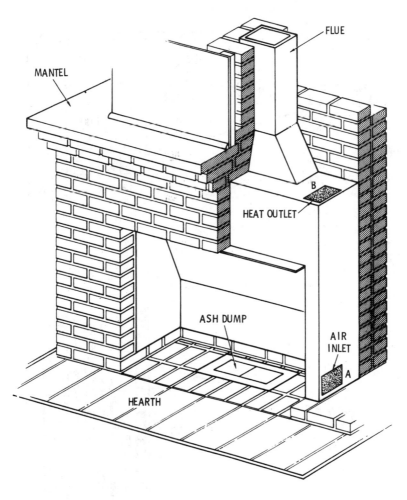

Fig. 9-17. Cutaway sketch of fireplace using a metal form.

To determine whether the fireplace opening is in correct proportion to the flue area, hold a piece of sheet metal across the top of the fireplace opening and then gradually lower it, making the opening smaller until smoke does not come into the room. Mark the lower edge of the metal on the sides of the fireplace.

The opening may then be reduced by building in a metal shield

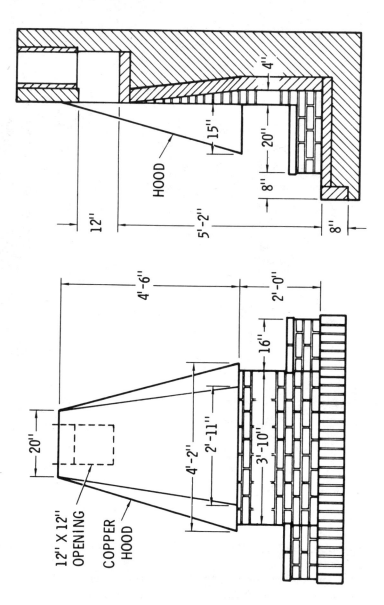

Fig. 9-18. A hood projecting out from the wall carries flue gases up through chimney.

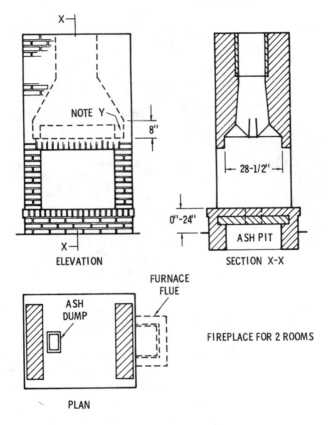

Fig. 9-19. A two-sided fireplace which is ideal for a room divider.

or hood across the top of the fireplace so that its lower edge is at the marks made during the test. Trouble with smoky fireplaces can also usually be remedied by increasing the height of the flue.

Uncemented flue-lining joints cause smoke to penetrate the flue joints and descend out of the fireplace. The best remedy is to tear out the chimney and join linings properly. When this is impractical, try the alternative method described in the following paragraph.

Where flue joints are uncemented and mortar in surrounding brickwork has disintegrated, there is often a leakage of air in the

211

Fig. 9-20. A grate is used to hold logs above the base, allowing air to move under and through the burning wood.

chimney, causing poor draft. This prevents the stack from exerting the draft possibilities which its height would normally ensure.

Another cause of poor draft is wind being deflected down the chimney. The surroundings of a home may have a marked bearing on fireplace performance. Thus, for example, if the home is located at the foot of a bluff or hill, or if there are high trees close at hand, the result may be to deflect wind down the chimney in heavy gusts. A most common and efficient method of dealing with this type of difficulty is to provide a hood on the chimney top.

Carrying the flue lining a few inches above the brickwork with a bevel of cement around it can also be used as a means of promoting a clean exit of smoke from the chimney flue. This will effectively prevent wind eddies. The cement bevel also causes moisture to drain from the top and prevents frost troubles between lining and masonry.

212

CHAPTER 10

Structural Clay Tile and Glass Block

Structural clay tile, also known as *hollow tile* (because it is hollow), is not nearly as popular as it once was (particularly during World War II), having been largely replaced by block, which is cheaper. Originally it came into popular use because of its ability to replace brick in some situations: It could be used for building walls, yet weighed only about one-quarter of what brick does.

A companion product, structural glazed facing tile, is indeed still very popular (more details later). If you want to know who makes plain structural clay tile, contact the Brick Institute of America in Washington, D.C.

STRUCTURAL CLAY TILE

Structural clay tile may be made of burned clay shale, the same as brick. It is made by forcing clay through special dies and then cutting it to various lengths. The hollow spaces are called *cells*, and the outside is called the *shell*. The inside partitions between hollow sections are called the *web*. The shell is at least ¾" thick, and the web is ½" thick. Structural clay tile is available in two basic strengths, two basic quality classes, and a number of facing finishes.

Load-bearing tile can be a direct substitute for load-bearing brick construction. Load-bearing tile is available in two qualities: grades LB and LBX. Grade LB is for masonry construction, not exposed to the weather, and requiring an outside coating at least 3" thick. Facing may be plaster, stucco, or a tier of regular brick. Grade LBX may be exposed to weathering without any facing layer. Load-bearing tile is used for backing up load-bearing brick walls or for interior partition walls. It is light in weight and has good sound- and heat-insulation qualities. When laid vertically, it may carry utility services, such as electrical conduit, gas and water pipes, telephone wires, and even heat and air-conditioning ducts. The outer scoring permits plastering directly to the tile without the need for furring.

Structural clay tiles, where not exposed, may be scored on all four sides. Where they are to be exposed, they are available in several surface finishes. Tile with one side smooth has the general appearance of brick inasmuch as they are made of the same material and fired in the same way. Type FTX is defect-free and has a smooth surface that is easy to clean. It is also called special-duty tile. Type FTS is inferior in surface quality but good for general-purpose use. Structural glazed facing tile has a hard glazed appearance. It may be ceramic-glazed, salt-glazed, or clay-coated. It is stainproof, very easily cleaned, and available in many colors.

Physical Characteristics of Tile

Structural clay tile comes in various sizes and shapes, as shown in Fig. 10-1. The sizes shown are nominal, based on the 4" module system, the same as brick. Actual sizes are slightly smaller to allow for mortar thickness.

The compressive strength of tile depends on the material used and the method of manufacture. It is also a function of the shell

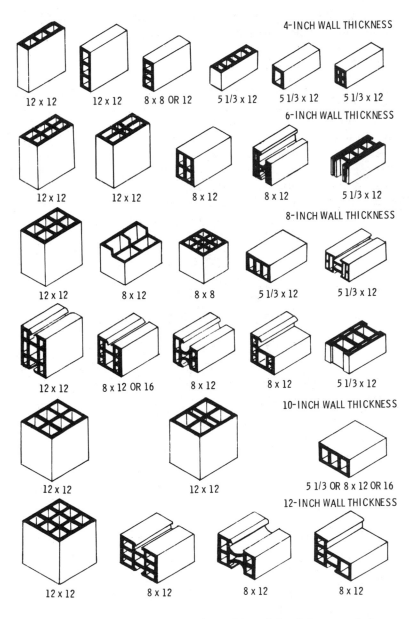

4-INCH WALL THICKNESS

12 x 12   12 x 12   8 x 8 OR 12   5 1/3 x 12   5 1/3 x 12   5 1/3 x 12

6-INCH WALL THICKNESS

12 x 12   12 x 12   8 x 12   8 x 12   5 1/3 x 12

8-INCH WALL THICKNESS

12 x 12   8 x 12   8 x 8   5 1/3 x 12   5 1/3 x 12

12 x 12   8 x 12 OR 16   8 x 12   8 x 12   5 1/3 x 12

10-INCH WALL THICKNESS

12 x 12   12 x 12   5 1/3 OR 8 x 12 OR 16

12-INCH WALL THICKNESS

12 x 12   8 x 12   8 x 12   8 x 12

Fig. 10-1. Structural clay tile comes in a wide variety of shapes and sizes.

and web thickness. The tensile strength, of course, is much less, probably less than 10% of the compressive strength.

Building with Hollow Clay Tile

Hollow tile may be laid horizontally or vertically and is constructed with ½" mortar joints. There is no real structural difference between the two styles. Laying them with the hollow cores horizontal is a little easier. Laid vertically, the hollow cores may be used for piping in the wall. In either case, the tiles of each course should be staggered, or overlap the course below and above, for best bonding, as for brick.

Fig. 10-2 shows the beginning corner using tile with overlapping bond. The first course must be laid on a level foundation of

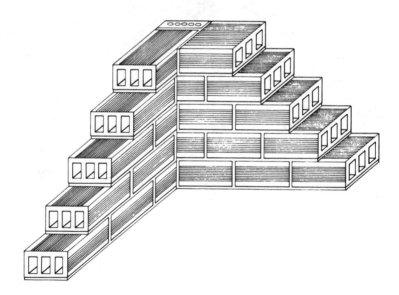

Fig. 10-2. How a corner is started with hollow tile.

concrete and with a 1"-thick mortar bed. Use a spirit level frequently to be sure the first course is perfectly horizontal.

Fig. 10-3 shows three steps in laying tile with a brick facing. The first course, on the foundation, consists of all header brick.

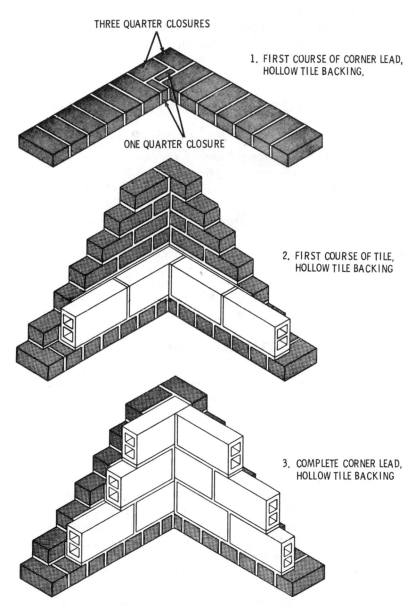

THREE QUARTER CLOSURES

1. FIRST COURSE OF CORNER LEAD, HOLLOW TILE BACKING.

ONE QUARTER CLOSURE

2. FIRST COURSE OF TILE, HOLLOW TILE BACKING

3. COMPLETE CORNER LEAD, HOLLOW TILE BACKING

Fig. 10-3. Steps in the construction of a tile wall with a brick facing.

The facing tier of brick follows the standard procedure described in previous chapters. These sketches indicate how many fewer pieces of tile need to be handled when the larger sizes are used, as compared to a solid brick wall.

STRUCTURAL GLAZED FACING TILE

Structural glazed facing tile, or SGFT, is a close cousin of structural clay tile. It is made of pulverized clay, which is extruded into a variety of shapes, usually 8″ × 16″ and 5″ × 12″ nominal faces × 2″, 4″, 6″ and 8″ depths. After the units have been extruded, its face or faces are glazed and fired at over 2000° to form a ceramic surface that is integral to the tile.

SGFT comes in a wide variety of colors and has many uses where a colorful, low-maintenance material is desired, such as in halls, lockerrooms, kitchens, and laundry areas. It comes in a variety of designs. For more information you can contact the Brick Institute of America in Washington, D.C.

GLASS BLOCK

For a long time, the use of glass block was in decline, but in recent years there has been a resurgence of interest, probably because people are reevaluating all kinds of materials. Glass blocks are used in industrial and residential settings for walls of all kinds (Fig. 10-4). It should be noted, however, that glass block is not load-bearing; some other means of providing support must be put into play.

Glass blocks are hollow glass structures made of pressed glass from which part of the air has been evacuated. These blocks are hermetically sealed at the time of manufacture, leaving a sealed-in dead air space. The edges of the blocks are of a gritty texture, making them mortar-binding. Some types of blocks on the market have a flanged "key-lock" edge; others have a corrugated type of edge which is in contact with the mortar.

Glass blocks are available in a variety of shapes, sizes, and designs. There are squares, radials, and corners. The designs

Fig. 10-4. Glass block can be used well in industrial as well as residential settings.

include convex and concave ribs or flutes, arranged vertically, horizontally, or both. The ribs may be on the interior or the exterior of the block. There are also clear, smooth, transparent types, which permit full view through them. The most widely used types are the square blocks which are 5¾", 7¾", and 11¾" on two of their dimensions, and 3⅞" thick (Fig. 10-5).

The advantages of glass blocks are of course dependent on their unique construction. Among the advantages are:

1. Excellent temperature control because of the sealed-in dead air space.
2. Low condensation on the surfaces compared to regular glass windows—therefore regulating humidity.
3. They transmit light, thus making it possible to daylight interiors which would otherwise be dark or artificially illumi-

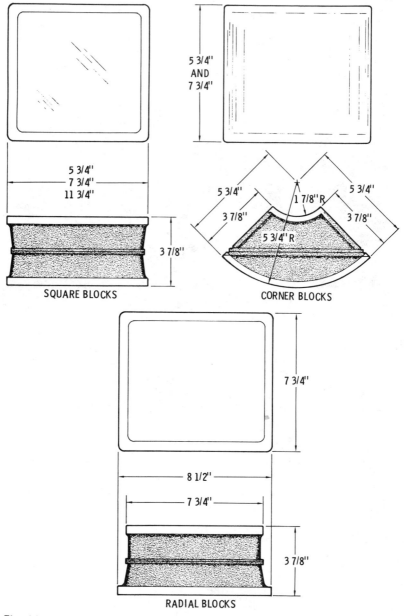

5 3/4"
AND
7 3/4"

5 3/4"
7 3/4"
11 3/4"

3 7/8"

SQUARE BLOCKS

5 3/4"
3 7/8"

1 7/8"R

5 3/4"R

5 3/4"
3 7/8"

CORNER BLOCKS

7 3/4"

8 1/2"

7 3/4"

3 7/8"

RADIAL BLOCKS

Fig. 10-5. Different sizes and shapes of installing glass block.

nated. They can also be used to control the direction of light transmission.

4. They are easy to maintain and easy to replace.
5. They provide privacy.
6. They reduce sound transmission.

These advantages suggest the use of glass block in industrial and office buildings which are completely air-controlled both in summer and winter. Where open windows are never required, glass-block windows and walls allow light to enter, yet provide excellent insulation against heat transmission. In homes, glass block is favored for use in stairwells and alcoves, where daylight illumination is desired but open windows are not necessary.

Glass blocks may be set as panels in masonry, concrete, steel frames, and wood frames, both interior and exterior, using mortar joints. Ventilation units in the form of windows or louvers may be installed in the panels. Construction in masonry usually employs the chase method. Where this is not practicable, wall-anchor construction may be employed.

The chase method is shown in Fig. 10-6. Indents in the masonry lock the edges of the glass block panel in place. Where the chase method is not practical, anchors are used, as shown in Fig. 10-7. The most economical size and the most widely used type of block is the 7¾" square. The smaller and the large type of squares are chosen when they will make a better working scale possible.

In selecting the design, it is well to keep in mind the purpose for which the installation is being made. The most important considerations will be the appearance and the type of light transmission required. Where the type of light transmission is not important, the design selected will be guided by the choice of the consumer or contractor.

Blocks which do not control the direction of light are the so-called general-purpose blocks, which are widely used. Where it is important to direct the light either upward or downward, or through the glass, the "functional" block is used. This type is designed so as to direct the light properly. These vary in their effects, and their selection is best made by consulting the descriptions and recommendations in the catalogs of the manufacturers.

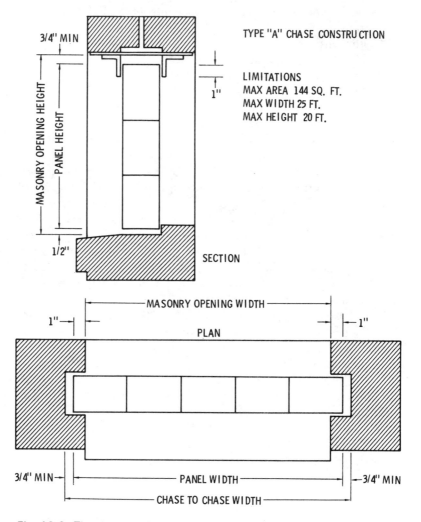

Fig. 10-6. The chase method of installing glass block.

Mortar mix used for the setting of glass blocks should have the following composition by dry volume: 1 part portland cement, 1 part lime, 4 to 6 parts sand. Mix the materials to a consistency as stiff as will permit good working. These materials should be drier than for ordinary clay brickwork. A metallic-stearate water-

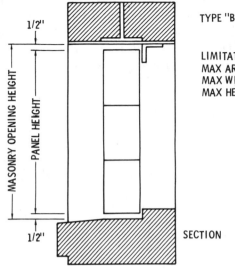

TYPE "B" WALL ANCHOR CONSTRUCTION

LIMITATIONS
MAX AREA 100 SQ. FT.
MAX WIDTH 10 FT.
MAX HEIGHT 10 FT.

SECTION

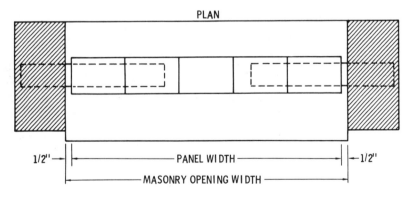

Fig. 10-7. The wall-anchor construction for glass block.

proofer is recommended by some manufacturers. The sand must be free from silt, clay, and loam in excess of 3% by weight as determined by decantations. Not more than 5% by weight will pass a No. 100 mesh sieve, and 100% will pass through a No. 8 mesh sieve.

223

Mortar thickness is generally only ¼″. Thus 5¾″, 7¾″, and 11¾″ square glass blocks are considered to have nominal sizes of 6″, 8″, and 12″ square. Mortar must be spread completely across all edge surfaces of the glass blocks to ensure watertightness and weathertightness. Carefully tool mortar joints and completely clean excess mortar off the glass.

CHAPTER 11

Brick Walks, Floors, and Terraces

Because of its weathering qualities, hard, well-burned brick was once used for roadways. Because of high labor cost in laying brick roads and the development of high-speed automatic road-building equipment, brick was replaced by asphalt and concrete. Because of its aesthetic values, brick is still used for many terraces, walks, and floors.

In some home design a goal is to match the interior with the exterior. This is especially true for families who believe in indoor-outdoor living. An example of this is shown in the sketch of Fig. 11-1, in which the terrace, the garden wall, and even planters extend into the house.

Thanks to the variety of brick available, there is virtually no limit to the tasteful architectural designs and uses to which brick may be put. Patterned walls go well with brick terraces (see Chapter 8 on various surfaces and patterns). Brick makes excel-

Fig. 11-1. Brick terrace, walls, and garden extend into the house to give that indoor-outdoor living look.

lent enclosures for trash cans (Fig. 11-2) or to conceal the condenser of a central air-conditioning system (Fig. 11-3). Brick planters (Fig. 11-4) make waist-high gardening easy. Improving the exterior with brick is easy to do, even for the homeowner, whether it is a simple grass edge for guiding the mower (Fig 11-5) or more elaborate jobs.

A sloping area can cause erosion of the topsoil from a prized tree. A brick retaining wall is easily put in place to straighten out the soil around the tree (Fig. 11-6). Brick terracing goes well with concrete open-grill block. Brick also fits well with various architectural styles, such as Early American and Spanish. Historic Georgetown is probably the best example of the early use of brick at the beginning of American history. "Old Town" in Albuquerque, New Mexico (Figs. 11-6 to 11-8) was originally settled

Fig. 11-2. The proper design of a trash-can enclosure made with brick can add to, rather than detract from, the garden architecture.

in 1706. By this time most of the brick walks have been replaced, but the original design has been maintained.

Brick is available in a variety of colors and textures, some of which are described in Chapter 8. Brick can be laid in many patterns, making overall style choices very large indeed. Fig. 11-9

Fig. 11-3. An enclosure for an outdoor central air-conditioning condenser. Open-wall pattern on the front brick wall permits good air circulation.

shows some of the patterns that can be formed with brick. Such choices make a brick terrace or walk preferable to the monotony of poured concrete.

TYPES OF BRICK FOR WALKS

Since walks and terraces must bear people, it is wise not to select brick with too rough a texture. A smooth texture is preferable, particularly if rolling equipment such as a mower or wheel-

228

Fig. 11-4. Properly designed brick walls make good earth retainers and planter boxes.

barrow is to be used occasionally. Brick must stand up against the elements of the area. Use only very hard, well-burned brick if weather is severe. Avoid rough seconds or textures, which can collect drops of water that may freeze and cause cracking.

For laying any pattern other than the stacked pattern (lower-right corner of Fig. 11-9), the brick size to order depends on whether you intend to mortar the joints or lay the bricks edge to edge. Nominal sizes of brick, as used for laying up a wall, allow for a ½" mortar thickness between bricks; however, brick is also available in actual sizes which follow the 4" module system. In actual sizes, the patterns shown can be laid without mortar joints. This is obvious when you note how many patterns require laying the 4" widths against the 8" lengths. Brick cannot be laid dry and tight unless the width is actually one-half the length. When laying with mortar, the same brick as used on walls will give the ratio of two half-widths to one length, as the mortar makes up the differ-

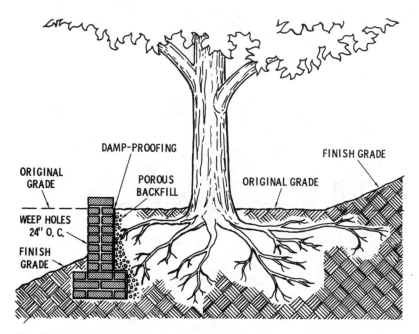

DAMP-PROOFING

FINISH GRADE

ORIGINAL
GRADE

POROUS
BACKFILL

ORIGINAL GRADE

WEEP HOLES
24' O. C.

FINISH
GRADE

Fig. 11-5. A brick retaining wall is used here to level the soil around tree and to prevent erosion.

ence. Some "pavers" can be bought, which have actual dimensions of 4″ × 8″ and are from 1½″ to 2¼″ thick. Also available is larger square-shaped, fired-clay brick.

INTERIOR BRICK FLOORS

Brick is frequently used in some residential and other structures for floors, where it fits the decor of the building style. Brick flooring for the den and some other rooms (sometimes all the rooms) fits well with ranch, Spanish, and adobe styling.

Although it is not required to stand up under severe weather, interior brick should be smooth and hard and should be carefully laid to form a perfectly level floor. Solid colors predominate, but patterns and variations in color are frequently appropriate. Textured brick is not recommended. In both interior floors and exte-

Fig. 11-6. New brick replaces old in sidewalks built in 1706.

Fig. 11-7. New brick replacing old steps. Old brick in upper portion of illustration was not mortared in place; notice signs of shifting.

Fig. 11-8. A very old brick walk laid without mortar. Notice the joint separation in some areas.

rior walks, brick may be laid with mortar joints or may be mortarless. Mortarless floors are less expensive to place and faster to complete.

Altogether, three layers are involved in laying interior brick floors:

1. The *base*—the principal support. It may be a concrete slab or well-tamped earth.
2. The *cushion*—a layer of sand about 2″ thick to facilitate leveling and placing of the brick.
3. The *brick*.

The best base is a concrete slab poured as part of the footing at the time the home is built. All requirements of good slab construction apply. Leveling the surface of the concrete is not as important as when it is used for a direct tile or carpeted floor since the sand cushion can be adjusted to compensate for any unevenness. The slab should include reinforcing rods and, in cold climates, a layer of insulation between the slab and the earth support.

In mild climates, and if the base earth is hard when dry and has little or no organic material, the base may be the earth itself. It

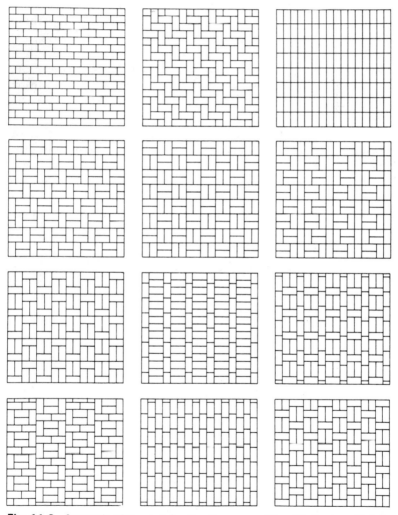

Fig. 11-9. A sample of the variety of patterns possible with the use of brick.

233

must be heavily tamped with the blunt end of a 4 × 4 or a tamping tool. The cushion of sand must be treated to reduce air spaces between sand granules. This is done by floating with water or tamping. Draw a straight board across in all directions to obtain a perfectly level surface. Any unevenness resulting from walking on it while laying the brick can be adjusted by hand at the time the brick is laid.

When considering a pattern with brick, keep in mind that there is less shifting if a herringbone or overlapping bond pattern is used, as against the straight stack pattern. Mortarless brick is laid edge to edge with the least possible space between bricks. As mentioned before, be sure actual dimensions of 4″ × 8″ brick are used. After the brick is laid, sweep very fine sand over the brick surface to fill any possible cracks.

If a mortared brick floor is used, the cushion should also be of mortar. However, the cushion is not poured all at one time. It is placed as a mortar bed for the brick as each brick is laid along with the edge mortar. The mortar cushion can be over a concrete base or an earth base. If over an earth base, it is wise to add strength by reinforcing the brick and the cushion with ¼″ steel rods.

Carefully wipe off any excess mortar from the surface of the brick as you work. Strike mortar joints flush. Use a rough rag for wiping. To increase the density of the mortar joints, they should be tooled. The height of the mortar should be the height of the brick.

EXTERIOR WALKS AND TERRACES

Because exterior walks and terraces are exposed to moisture and weather, it is important that the brick selected have low absorption qualities and be extra hard and well-burned. It should meet the requirements for grade SW of ASTM standards for facing brick. A good grade of hard brick naturally has the qualities for long life and good weathering and is an excellent choice for walk material where the style is appropriate.

Basically, exterior walks and terraces follow the same construction as for interior floors, requiring a base, a cushion, and the final

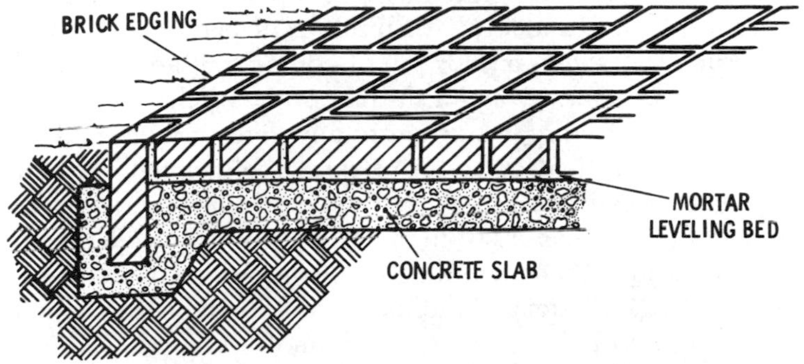

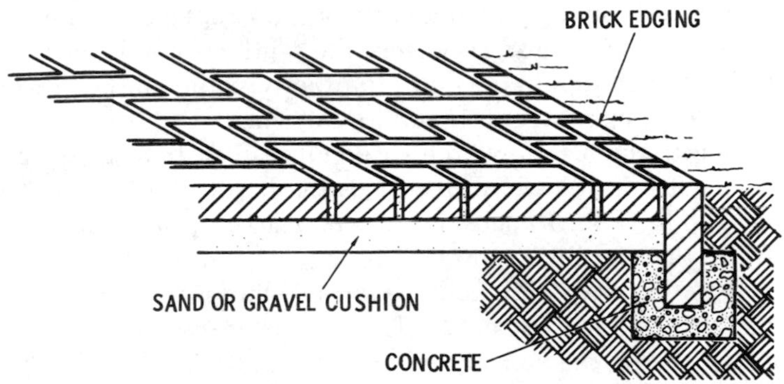

Fig. 11-10. Mortared brick walk with earth base and with concrete base.

brick layer, but with a few other considerations for weathering. The base is well-tamped earth or a concrete slab. The cushion is a layer of sand for mortarless brick or a thin bed of mortar for mortared brick. The sketch of Fig. 11-10 shows both in a cutaway view.

The edging shown in Fig. 11-10 is a most important part of walk construction if a carefree walk is to be achieved. If the earth is used as a base, dig a trench along both edges of the walk and pour concrete for a footing. Lay a row of rowlock brick on edge after you pour the concrete. Of course, it will be necessary to measure accurately the distance between edges to be sure that the width will accept the desired pattern and that the bricks will come out even. With a concrete base, a deeper footing is dug for the edges of the slab and a row of bricks is placed on edge. The walk in Fig. 11-10 shows a poor job of edging, with evidence of the edging breaking away and a shifting of the brick.

Good drainage is very important to any kind of work laid on the ground, and particularly so for bricks. In cold climates, water collecting between bricks can freeze and crack the edges or push bricks out of place. In areas of high humidity and/or with a high water table, water can accumulate for long periods of time and cause the growth of fungi and molds. Walks should be sloped or a slight crown provided to permit water runoff. A slope of $\frac{1}{8}$" to $\frac{1}{4}$" per foot is adequate. For good drainage from underneath, sandy soil will provide the necessary treatment. Where the soil is hard, dig away approximately 2 ft. and replace it with gravel. However, where gravel is used, the sand cushion layer for mortarless laying should be omitted; otherwise the sand will work its way down into the gravel and lose its value.

Construction follows the directions given before for interior floors: Tamp the soil well (this is the base). Use a bed of well-leveled sand for the cushion for mortarless brick; for mortared brick, the cushion is mortar.

If the base is sandy and a good leveling job can be done, the sand cushion may be omitted for a mortarless layer of brick. To reduce labor, the mortar bed may be omitted. While the durability of the final job is less, the reduction is not in proportion to the reduction in labor. Type S mortar should be used when there is a concrete base. This consists of 1 part portland cement, $\frac{1}{2}$ part lime, and $4\frac{1}{2}$ parts sand. On an earth base, use Type M Mortar, consisting of 1 part portland cement, $\frac{1}{4}$ part lime, and 3 parts sand. Portland masonry cement with the proper amount of lime already added may be purchased.

Mortaring can be done by hand, of course. Two methods which reduce the amount of labor are frequently used. These are the dry-mix method and the pourable-grout method. In the dry-mix method, place the brick into position, leaving a $\frac{1}{2}$" space between each brick. Make a dry mix of 1 part portland cement and 3 or 4 parts of fine dry sand. Sweep the dry mix into the spaces between the bricks. Spray with a fine mist of water until the walk is damp. The water will seep into the mix and start hydration. Keep the walk damp for two or three days for good curing strength. It is important that the mix be swept into the cracks and completely off the top surface of the brick. Otherwise, the brick may stain when the cement is watered.

In the pourable-grout method, a pourable grout is made of cement, sand, and water. Use slightly higher amounts of sand than the dry-mix method. Add water to make a soft and loose grout which can be poured into the spaces between the brick, from the lip of a bucket, or the spout of a large watering can.

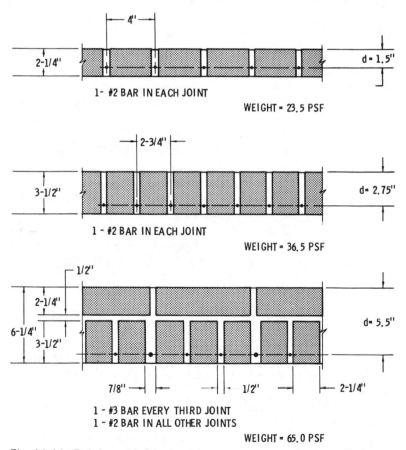

Fig. 11-11. Reinforced brick may be necessary over culverts or ditches or where there is no firm base..

Reinforced Brick

When a walk must span a culvert or small ditch, the brick should be reinforced or it will soon break down under the weight of traffic. It must be supported on a reinforced-concrete foundation. Wood and sheet steel spans can be used if the brick is reinforced. Fig. 11-11 shows three methods of reinforcement, depending on the thickness of the brick. The joints should be well tooled to make sure that they are dense and that the mortar fully surrounds the reinforcing rods.

Steps

Where the path of a walk runs into a rise or drop in the earth, the best way to handle the change in elevation is by means of steps. Steps of brick will match the rest of the walk and can be adjusted to fit the slope. Fig. 11-12 shows how the same number of steps can handle two different angles of slope by the way in which the brick is laid. In Fig. 11-12A, a smaller angle of rise uses brick on its face; in Fig. 11-12B, setting the brick on edge, row-lock style, increases the rise angle.

Because an earth base and sand cushion can be too easily affected by traffic and erosion in heavy rain, it is best to make the base of poured concrete to maintain a firm set. After determining tread and riser dimensions from brick sizes, make wooden forms for the poured concrete. Include reinforcing rods, which are indicated by the dotted lines in the sketch of Fig. 11-12.

RETAINING WALLS

Brick walls make excellent retaining walls, whether for holding dirt for planting or for terracing a sloping garden. Construction is similar to that of a wall for structures, with a few precautions added.

An earth retaining wall is subject to lateral pressures from the earth it is holding. It is also exposed to a considerable amount of moisture. Depending on the amount of earth fill the wall is to hold, extra care is needed to make sure the wall has high lateral strength. All joints must be well mortared over the entire surface

238

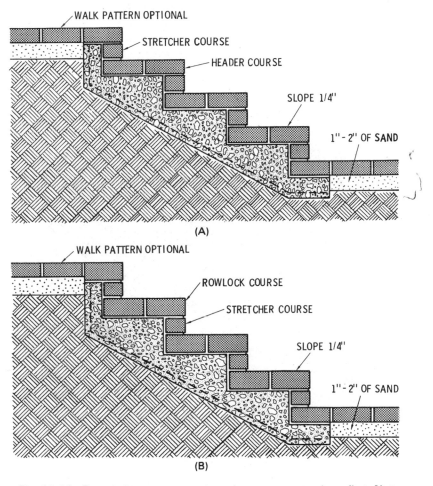

WALK PATTERN OPTIONAL

STRETCHER COURSE

HEADER COURSE

SLOPE 1/4"

1" - 2" OF SAND

(A)

WALK PATTERN OPTIONAL

ROWLOCK COURSE

STRETCHER COURSE

SLOPE 1/4"

1" - 2" OF SAND

(B)

Fig. 11-12. For good step support, pour the concrete step base first. Note how brick can be laid to change the degree of rise.

of the bricks. If a large amount of earth fill is to be retained by the wall, add steel reinforcing rods.

Because stains and efflorescence can develop on brick when exposed to moisture over long periods of time, care must be used to prevent the moisture of the earth from penetrating the brick. Fig. 11-13 shows the details of a retaining wall. Weep holes in the wall, made by leaving the mortar out of some vertical joints, allow moisture to drain through the wall to the outside and reduce the absorption of water by the brick.

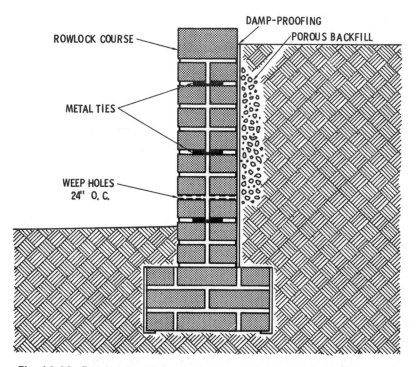

Fig. 11-13. Details of retaining wall illustrated in Fig. 11-5. Tar paper, rock backfill, and weep holes are precautions against moisture absorption by the brick.

Fig. 11-14 is an example of a typical brick planter construction. Coarse gravel below and above the foundation, connected by a pipe through the foundation, drains away moisture settling at the bottom of the planter. In addition, tar paper between the dirt and brick wall prevents seepage into the brick.

Cleaning Up

If a mortared brick floor or walk is laid, it is almost impossible to prevent mortar droppings from falling onto the face of the

240

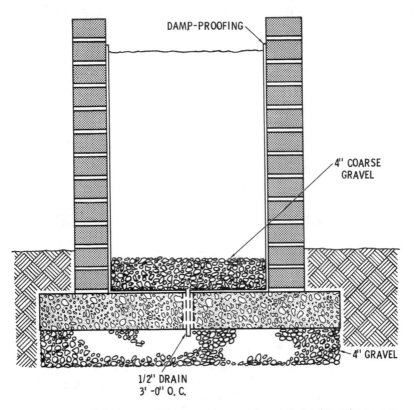

DAMP-PROOFING

4" COARSE GRAVEL

4" GRAVEL

1/2" DRAIN
3' -0" O. C.

Fig. 11-14. Brick planter with tar paper, gravel, and drain pipe at bottom to reduce moisture absorption.

brick. It is essential to work with care to reduce these droppings to a minimum. Most of them can be cleaned off by wiping the brick with a burlap cloth, but this must be done while the mortar is still damp.

Once cement has hardened, it cannot be removed with water. There is only one solution that dissolves dried cement, and that is muriatic acid (commercial grade hydrochloric acid). But acid is not the easy solution. Acid is extremely harmful to skin and clothing, and especially dangerous to the eyes. Acid and its fumes can corrode metals such as aluminum and galvanized steel. It is injurious to plant life. It should be handled only by the most expe-

241

rienced mason using rubber gloves, goggles, and plenty of ventilation.

There are some brand-name cleaning solutions which frequently are inhibited acids or are self-neutralized. These are less harmful than muriatic acid, but more costly. Before applying acid, a fine spray of water should be applied to the brick. The layer of water reduces the absorption of acid into the brick, especially into the mortar joints, reducing the possibility of "acid burn" on the brick. Scrub the acid onto the brick, using a stiff brush and working only a small area at a time. Thoroughly rinse off all acid afterward.

BRICK-FLOOR FINISHING

Mortared floors may be given a perfectly smooth finish by grinding with a terrazzo grinder. However, floor grinding should be anticipated before the floor is laid as certain requirements are essential in advance. Brick to be ground should be of the high-density type, hard-burned and nonporous. On softer brick, the grinding will merely expose the less dense clay under the surface, which may have small air bubbles that can absorb water. In addition, the sand used for the mortar should be extra fine. Coarse sand granules will be dislodged by the grinder and will tend to scratch the brick surface.

Brick should be allowed to set for a few days to make sure the mortar has begun to hydrate well. Otherwise, the grinder may dislodge some of the brick. A wax coating is frequently used on indoor brick floors to reduce the absorption of water from mopping. Wax should not be used on outdoor walks because it may tend to hold water under the wax, which can freeze on cold days.

One of the better treatments for brick is a silicone coating. Silicone has a long life and is a good sealer against water. Use masonry silicone sealer, which is available in two types: solvent-based and water-based. The solvent-based silicone dries more quickly, and waxing, if desired, may be applied sooner.

CHAPTER 12

Brick Repair and Maintenance

The need for repairs and cleaning on new brick masonry is reduced when the original construction is done in a careful manner and certain precautions are observed. The bricklayer craftsman should take pride in the work he is doing. A clean job with well-tooled and clean joints in the first place means less cleanup afterward and is less costly in the long run.

Piles of bricks waiting for use should be stacked off of the ground on plattens or rows of $2'' \times 4''$ lumber. This will keep the lower layer of bricks from absorbing moisture from the ground. If the brick is not to be used immediately, it should be covered with a tarpaulin to keep off rain and dust.

As a wall is laid up, debris of mortar will naturally fall to the ground. It can be kept away from the base of the wall by placing a layer of sand or a vinyl sheet at the base. Scaffolding should not abut the wall. Mortar droppings should be allowed to fall to the ground. At the end of the day, the walk planks near the wall should be turned back to prevent rain and wind from blowing collected dust onto the wall.

As the wall work progresses, excess mortar should be struck even with the face of the wall to prevent mortar from running down the face of the brick. Fill taut-line holes immediately after the pins are removed. Tool joints after the mortar is thumbprint dry. Tool firmly, moving in one direction, to prevent forming air pockets. After tooling, cut off tailings and brush excess mortar from the brick.

When the wall is finished, inspect it for loose pieces of mortar. Remove them with a chisel, if dry, and brush away loose particles. Use a hose to wash off cement dust.

CLEANING NEW BRICK

After cleaning the brick as mentioned before, some staining may appear. Stains are cleaned as follows:

Mortar stains are removed with hydrochloric acid prepared by mixing 1 part commercial muriatic acid with 9 parts water. Pour the acid into the water. Before applying the acid, soak the surface thoroughly with water to prevent the mortar stain from being drawn into the pores of the brick.

The acid solution is applied with a long-handled, stiff-fiber brush. Proper precautions must be taken to prevent the acid from getting on hands, arms, and clothing. Goggles are worn to protect the eyes. An area of 15 to 20 sq. ft. is scrubbed with acid and then immediately washed down with clear water. All acid must be removed before it can attack the mortar joint. Door and window frames must be protected from the acid.

Cleaning Off Efflorescence

Efflorescence is a white deposit of soluble salts frequently appearing on the surface of brick walls. These soluble salts are contained in the brick. Water penetrating the wall dissolves out the salts, and when the water evaporates, the salt remains. Efflorescence cannot occur unless both water and the salts are present. The proper selection of brick and a dry wall will keep efflorescence to a minimum. It may be removed, however, with the acid solution recommended for cleaning new walls. Acid should be used only after it has been determined that scrubbing with water and stiff brushes will not remove the efflorescence.

CLEANING OLD BRICK

Sandblasting, steam with water jets, and the use of cleaning compounds are the principal methods of cleaning old brick masonry. The process used depends on the materials used in the wall and the nature of the stain. Many cleaning compounds that have no effect on brick will damage mortar. Rough-textured brick is more difficult to clean than smooth-textured brick. Often, brick cannot be cleaned without removing part of the brick itself, thus changing the appearance of the wall.

Sandblasting

This method consists of blowing hard sand through a nozzle against the surface to be cleaned. Compressed air forces the sand through the nozzle. A layer of the surface is removed to the depth required to remove the stain. This is a disadvantage in that the surface is given a rough texture which can collect soot and dust. Sandblasting usually cuts deeply into the mortar joints, and it is often necessary to repoint them. After the sandblasting has been completed, it is advisable to apply a transparent waterproofing paint to the surface to help prevent soiling of the wall by soot and dust. Sandblasting is never done on glazed surfaces. A canvas screen placed around the scaffold used for sandblasting will make it possible to salvage most of the sand.

Steam with Water Jets

Cleaning by this method is accomplished by projecting a finely divided spray of steam and water at a high velocity against the surface to be cleaned. Grime is removed effectively without

245

changing the texture of the surface, which gives it an advantage over sandblasting.

The steam may be obtained from a portable truck-mounted boiler. The pressure should be from 140 to 150 psi, and about 12 boiler horsepower per cleaning nozzle is required. The velocity with which the steam and water spray hits the wall is more important than the volume of spray used.

A garden hose may be used to carry the water to the cleaning nozzle. Another garden hose supplies rinsing water. The operator experiments with the cleaning nozzle in order to determine the best angle and distance from the wall to hold the nozzle. The steam and water valves may also be regulated until the most effective spray is obtained. No more than a 2-ft.-sq. area should be cleaned at one time. The cleaning should be done by passing the nozzle back and forth over the area and then rinsing it immediately with clean water before moving to the next space.

Sodium carbonate, sodium bicarbonate, or trisodium phosphate may be added to the cleaning water entering the nozzle to aid in the cleaning action. The amount of salt remaining can be reduced considerably by washing the surface down with water before and after cleaning.

Hardened deposits that cannot be removed by steam cleaning should be removed with steel scrapers or wire brushes. Care must be taken not to cut into the surface. After the deposit has been removed, the surface should be washed down with water and steam cleaned.

Cleaning Compounds

There are a number of cleaning compounds that may be used, depending on the stain to be removed. Most cleaning compounds contain material that will appear as efflorescence if allowed to penetrate the surface. This may be prevented if the surface to be cleaned is thoroughly wetted first. Whitewash, calcimine, and cold-water paints may be removed by the use of a solution of 1 part hydrochloric acid to 5 parts water. Fiber brushes are used to scrub the surface vigorously with the solution while the solution is still foaming. When the coating has been removed, the wall must be washed down with clean water until the acid is completely removed.

PAINTING

A large selection of paint colors is available for painting brick, but it should be noted that painting is for appearance only—it won't improve resistance to weather. Cement paint is one kind you can use. This is cementitious in nature, having a small amount of portland cement as part of the ingredients. To make sure of proper adhesion to the brick, it is necessary that the brick be clean from any dust. Wash down the brick with a stream of water and let dry before painting.

Latex paints have a flat finish. They dry quickly and have a tough finish that is impregnable to water and oil and also stain-proof. Latex paint must be applied to brick that has an etched finish. It will not stick to painted brick.

WATERPROOFING MATERIALS

There are numerous proprietary mixtures on the market, but they appear to be of little or no benefit as waterproofing when applied to walls that leak badly, according to the results of tests at the National Bureau of Standards. Where wall leakage was through very fine cracks between the joints and the masonry units, some colorless waterproofers were effective for a period of a year or two, but after weathering, their effectiveness as waterproofers became considerably lessened. However, there are colorless waterproofers on the market for which fine performance records are cited. Therefore, it is recommended that before using such a material its performance in similar jobs in the area be investigated carefully.

Much publicity has recently been given the so-called "silicone" colorless water repellents (Fig. 12-1). Their indiscriminate use, however, should be cautioned against. There is strong evidence that if such silicones are applied to walls that are not properly built in terms of tight mortar joints, good flashing, etc., and where water can enter the wall from the back side due to condensation, the silicone treatment can actually cause damage to the wall.

If soluble salts are present in the wall, any water entering through defective joints, flashing, or from the back side will dis-

Fig. 12-1. Clear waterproofers being used on brick.

solve those salts. This salt solution will travel toward the face of the wall until stopped by the silicone protective layer, where the water will evaporate, leaving the salt crystals. If this process continues, the pressures eventually developed by the formation of these salt crystals can be sufficient to cause serious slabbing of the brick. Therefore, while silicone water repellents may be effective in stopping the appearance of efflorescence on the exterior surfaces of masonry walls, their improper use may eventually cause disintegration of the brick or tile units.

TUCK POINTING

As mortar dries, some shrinkage occurs. This forms small cracks in which water can collect. Because of the need to make mortar easily workable, the addition of lime and the ratio of water to cement result in mortar being somewhat weaker than concrete. In time, freezing and thawing action can cause deterioration and some of the mortar must be replaced. This is called *tuck pointing.*

Tuck pointing consists of cutting out all loose and disintegrated mortar to a depth of at least ½″ and replacing it with new mortar. If leakage is to be stopped, all the mortar in the affected area should be cut out and new mortar placed. Tuck pointing done as routine maintenance requires the removal of the defective mortar only.

All dust and loose material should be removed by a brush or by means of a water jet after the cutting has been completed. A chisel with a cutting edge about ½″ wide is suitable for cutting. If water is used in cleaning the joints, no further wetting is required. If not, the surface of the joint must be moistened.

Mortar for Tuck Pointing

The mortar to be used for tuck pointing should be portland-cement-lime, prehydrated Type S mortar, or prehydrated prepared mortar made from Type II masonry cement. The prehydration of mortar greatly reduces the amount of shrinkage. The procedure for prehydrating mortar is as follows: The dry ingredients for the mortar are mixed with just enough water to produce a damp mass of such consistency that it will retain its form when compressed into a ball with the hands. The mortar should then be allowed to stand for at least 1 hour and not more than 2 hours. After this, it is mixed with the amount of water required to produce a stiff but workable consistency.

Filling the Joint

Sufficient time should be allowed for absorption of the moisture used in preparing the joint before the joint is filled with mortar. Filling the joint with mortar is called *repointing* and is done with a pointing trowel. The prehydrated mortar that has been prepared as above is packed tightly into the joint in thin layers about ¼″ thick and finished to a smooth concave surface with a pointing tool. To reduce the possibility of forming air pockets, the mortar is pushed into the joint with a forward motion in one direction from a starting point.

CHAPTER 13

Brick Patio Projects

Building with brick is really quite easy. A number of garden projects, involving little material and time, make excellent training lessons for the apprentice brickmason or easy projects for the homeowner. This chapter describes a number of projects, easy enough to complete in one weekend, for the handyman who wants to do-it-yourself or for the beginner bricklayer to pick up a few extra dollars during time off from his regular job.

One aspect of the projects described here that makes them easier is the use of premixed concrete and mortar. Nearly all cities have one brand or another of packaged dry ingredients for the mixing of concrete or mortar. The packages include the right amount of cement and other ingredients, requiring only the addition of water to make concrete or mortar. Instructions for adding the water are given on the bag. They are available in various weights and sizes. Brand names may vary from city to city, depending on the local supplier. One national brand available almost everywhere is Sakrete.

PREMIXED MATERIALS

Packaged concrete as a dry mix can be purchased in three basic forms and in several sizes:

Concrete. The ingredients are cement, sand, and larger aggregates. This is the mixture for the base of many of the projects in this chapter. It can be obtained in 90-lb. and 25-lb. sacks.

Sand Mix. A mixture of sand and cement for use on shallow bases and for repairs. It can be purchased in 80-lb. and 25-lb. sacks and 11-lb. packages.

Mortar. A mixture of cement, lime, and sand for use as a brick mortar. It is avialable in the same sizes as the sand mix.

A FLAGSTONE WALL

Flagstone is a natural rock avialable in many shapes and sizes, from large flat pieces for walks to smaller and thicker pieces approximately the size and shape of bricks. Its naturally rough texture and size variations give it a rustic appearance, especially suited for use in garden architecture.

Figs. 13-1 to 13-9 show a flagstone wall being built as a retaining wall after dirt has been dug away from a high-rising small hill. This will be disregarded, and only the construction of the wall considered. (Note the concrete block for the first two courses, which later will be under water.)

As is necessary for any brick wall, a firm foundation on which to build is the order of first importance. Dig out a trench about 5″ deep and 10″ to 12″ wide in areas of mild climate, or down to the frost line in cold climates. The trench will be for a concrete base. If the soil is firm enough to be dug out with straight sides, concrete can be poured directly into the trench. If the soil is quite sandy and straight sides cannot be maintained, it will be necessary to install board forms to hold the concrete.

Pour a sack of Sakrete (or other packaged concrete) mix into a wheelbarrow and add the prescribed amount of water. Using a garden hoe, thoroughly mix the dry ingredients and water. Pour the concrete into the trench to about the halfway point. Lay a couple of ½″ steel rods into the trench for reinforcing, and fill the rest of the trench with concrete. This may require mixing several wheelbarrows of concrete, depending on the length and depth of the wall. Drive wooden stakes into the ground at each end of the base. Tie a heavy string between the stakes. Use a spirit level and adjust the taut line level. With the taut line as a guide, trowel the top of the concrete until it is perfectly level, or equidistant from all parts of the line. Allow about 24 hours for the concrete to take a good set before building up the wall.

Flagstone is quite irregular as to texture, size, and shape. To lay up a wall of flagstone with good overlap bonding, some uniformity of size is needed, and high spots should be removed. Use a

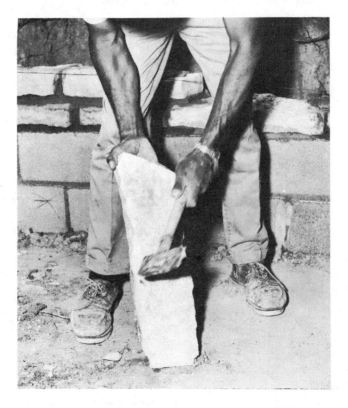

Fig. 13-1. A mason's hammer used to knock high spots off flagstones.

253

mason's hammer (Fig. 13-1) to knock off high spots and rough edges, just enough to prevent interference with good construction.

If the wall is a one-man job, it is best to mix only one sack of mortar, lay up the stones with the amount mixed, then mix another sack of mortar, etc. Pour the mortar mix into a wheelbar-

Fig. 13-2. Mix only as much premixed mortar mix as can be handled in 2 hours.

row and add the amount of water indicated on the sack. Mix the dry ingredients with the water, using a garden hoe, by a push-pull action (Fig.13-2). Push the material out from the center, and draw it back from the edges over into the center. A shovel may also be used, of course, but it is a little less handy.

Lay a 2″-thick layer of mortar across the top of the concrete base. Lay a row of flagstone in line from one end to the other, buttering one edge of each stone as you place it. Place a loose stone at each end of the wall, and reset the taut line between the two stones (Fig. 13-3). Fasten the line to sticks held by the stones. Be sure the line ends are the same distance above the base at each end. This becomes the guide for adjusting the mortared stones to level. Tap down each stone that is above the level of the line, using the back of the trowel.

Fig. 13-3. Reset the taut line for each course of stone. The end of the taut line is secured to the stone corners.

Succeeding courses of flagstone should be laid up like brick. Place about a 1″ layer of mortar on each of the corners, and place the corner stones. These will support loose stones temporarily for holding the reset taut line. Work from the ends to the center (Fig. 13-4). Tap each stone about level with the taut line (Fig. 13-5). Excess mortar squeezed out may be removed with the trowel and thrown onto the bed for the next stone (Fig 13-6).

For greater bonding strength and better appearance, try to maintain a running bond overlap, as one would with brick. Use the level frequently (Fig. 13-7) for horizontal positioning of each stone. Lay only about four or five stones at one time. As each course is completed, the taut line is raised to the next level. It must be rechecked each time with a level, or the ends must be carefully measured for equal distance from the base.

The mortar in the joints between the stones must not be allowed to set hard before tooling. About ½ hour after the mortar has been placed is a good time to tool, whether the wall has

Fig. 13-4. Laying a bed of mortar on the course just finished. The bed is generally thicker for flagstone than for brick.

Fig. 13-5. Tap stone into place and level to the taut guideline.

Fig. 13-6. Squeezed-out excess mortar is struck off and thrown onto the bed ahead.

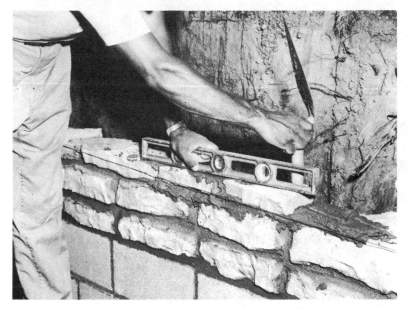

Fig. 13-7. Using a level frequently as stones are laid into position.

been completed or not. A slow worker may be tooling only three or four courses below the one being laid. Use a jointing tool or a rounded stick, and press it hard into the joints, pushing mortar in about ½ ″ behind the surface of the stones. Tooling increases the density of the mortar, making it more watertight, and improves the appearance. The tool is held at a slant as it is drawn across the mortar (Fig. 13-8). Push excess mortar feathers off with the tool.

As a finishing touch, brush the mortar joints with a stiff brush (Fig. 13-9). Use the brush to clean off small pieces of mortar from the face of the stones. If the mortar has hardened some, dampen the brush before using it.

A BRICK WALL

Chapter 11 described brick walks and terraces—the type of best-suited brick material, patterns, and construction. Shown here (Fig. 13-10) is the step-by-step procedure for building the

Fig. 13-8. A jointing tool or a rounded stick used to tool the mortar into a dense and neat joint.

Fig. 13-9. A stiff brush used to smooth the mortar after tooling.

259

Fig. 13-10. How well a brick walk can blend with stone in a garden or patio area.

small walk. The same procedure applies to a brick walk or patio of any size and pattern. This walk was built on a 2″ sand base, plus a 1″ concrete cushion, plus mortared brick. This calls for excavating the dirt for a depth of about 5½″ from the top level of the walk. The difference is the thickness of the brick.

After digging out the dirt, place form boards along the edges of the walk (Fig. 13-11). Make sure the board tops are level with

Fig. 13-11. Setting form boards for sidewalks.

the top of the walk surface and perfectly horizontal along the length of the walk. Parallel boards should be set so that one is slightly lower than the other for a slight tilt to the walk for water runoff. If the boards are to be removed later, ordinary 2″ × 4″ boards may be used. If they are to be left in place, as many prefer, use a wood or preservative-treated material which will not decay.

For building up a more secure edging, as recommended in Chapter 11, dig edge trenches deeper and pour concrete until the top of the concrete is about 4″ below the top of the form boards. Mix mortar and place a ½″ layer over the concrete. This will form the bed for the edge brick (Fig. 13-13). Put edge bricking in place and tap to exactly match the top edge of the board (Fig. 13-14). The brick is placed on edge, rowlock style. Be sure to butter the edge of the brick with mortar. Tool the mortar joints smooth. Allow 24 hours for the mortar to set before proceeding with the walk surface. Make up a leveling board like the one shown in Fig. 13-15. The bottom edge should be the right depth for the top of the sand base. Pour in sand and level it with the board.

Fig. 13-12. Concrete foundation to support edging brick for walk.

Fig. 13-13. Mortar bed spread over concrete foundation.

Fig. 13-14. Placing edging brick in mortar.

Fig. 13-15. Using a board to level the sand base.

Fig. 13-16. Dry-sand mix is poured over the sand base and leveled.

The next step is the laying of a concrete cushion. This may seem contrary to the method described in Chapter 11, but it is an alternative method which is also simpler. The secret of this method is the use of dry sand mix which is poured dry over the sand and leveled. This method is usable only in areas with considerable moisture in the ground since it depends on moisture seeping up through the sand base and penetrating the concrete dry mix to eventually hydrate and form a hard concrete cushion.

Remake the leveling board so that the bottom edge is only the depth of a brick, usually 3⅝″. Pour the dry sand mix over the sand base (Fig. 13-16) and level it carefully with the leveling board (sometimes called a *screed*). Having established the brick pattern desired, lay the brick onto the dry mix, allowing for space between each brick, from ⅜″ to ½″ (Fig. 13-17).

Mix dry mortar ingredients with water, but this time use about 30% more water than prescribed. The purpose is to make a mortar that is quite fluid. Pour the mortar into the spaces. Where the mortar does not fill completely, use the trowel to force it down

264

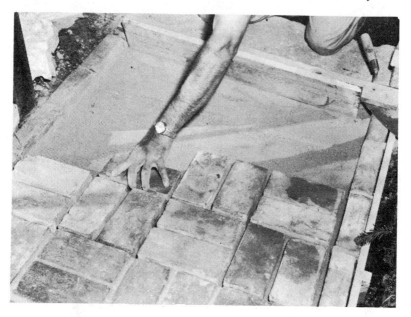

Fig. 13-17. Bricks are laid in place over the dry-sand mix.

(Fig. 13-18). When the mortar is about "thumbprint dry," use a jointing tool and work each of the joints smoothly (Fig. 13-19). Brush away all the loose mortar. If, after four days, it appears that some of the cement from the mortar is still on the brick, it should be removed before it stains the brick. The only solvent for cement is about a 10% solution of muriatic acid. Rinse it off thoroughly afterward. Use care in handling muriatic acid. The walk should be allowed to set undisturbed for 48 hours before using.

A CIRCULAR BARBECUE PIT

A barbecue pit in the center of the patio has some advantages (Fig. 13-20). It becomes the focal point for gatherings. Marshmallow toasters can gain access from any point around it. In the evening, when firewood is used, it can cast a pleasant light and provide some heat.

Fig. 13-18. Mortar is poured into the joint spaces between the bricks. A trowel is used to force mortar completely into all spaces.

Fig. 13-19. Tool mortar before joints harden.

Fig. 13-20. A circular pit in the center of the patio can be the focal point of outdoor gatherings.

It will be seen from the Figs. 13-21 to 13-29 that the concrete for the patio was laid some time during the middle of the construction of the barbecue pit. The patio may be built before, during, or after the pit is built. If built before, of course, the area for the pit should be left open.

The pit should be at least 50″ in diameter and dug to a depth of 19″ below the patio level (Fig. 13-21). Lay a 2″ bed of gravel, and drive a stake into the center. The stake will be used for height and radius measurements. Mix a small amount of dry sand mix with water, and place a 2″ layer over the gravel around the edges. While this is still wet, smooth the top and scribe a circle from the stake with a string and nail. This will establish an outer-diameter guide for laying the brick (Fig. 13-22). This will be the base for the buildup of a circular brick wall. Allow 24 hours for the concrete to set.

The first course of brick is set on edge (soldier style) on a ¾″ to 1″ bed of mortar. The bricks should be dampened before being mortared in place. Butter the edges with mortar, and fol

267

Fig. 13-21. The barbeque pit in the center of the patio is 19" deep and 50" in diameter, with a 2" layer of gravel in the bottom.

Fig. 13-22. A 2" thick layer of concrete in a circle will be used as the foundation for the brick barbecue pit. Notice the circle being scribed from the center stake.

low the circle around the base (Fig. 13-23). This course must be measured frequently with a level to make sure it is perfectly horizontal. It forms the base and guide for the remaining courses. Above the first course of rowlock brick, set another course, also rowlock style (on end). This should bring the construction level up to the level of the ground or patio surface. Two courses of brick above the patio floor should be enough. These are laid header style and with an overlapping pattern.

Lay a complete circle of brick dry to establish the number of bricks and the space between them. Remove about five of the bricks, and lay down a ½″ bed of mortar. Butter the ends of the bricks removed, and put them back in place. Be sure the joints between bricks are completely filled with mortar. The mortar joints will be wider at the outer edge than the inner. Figs. 13-24 to 13-28 show the steps. Note the straight stick in Fig. 13-25 used to adjust a newly laid brick to the level of the preceding one. Use a spirit level frequently. Use the handle of the trowel to tap bricks into position.

Fig. 13-23. Fire brick being mortared and set on end, which forms the circle for the barbecue pit.

269

Fig. 13-24. The top two courses of brick are laid as all headers with an overlap to improve bonding.

Fig. 13-25. Lay bricks on a bed of mortar, tapping them into place. A straight stick is used for alignment.

Fig. 13-26. Striking off excess mortar.

Fig. 13-27. Bricks must be placed carefully to make sure the circle is even.

271

Fig. 13-28. Make sure all joints are completely filled with mortar.

Allow about 15 minutes for the mortar to set, and then strike off excess mortar level with the brick (Fig. 13-29). This rule should be observed as you progress with the work. Do not wait until the entire job is done to smooth off excess mortar. Be sure to scrape off mortar that has fallen to the patio concrete.

GARDEN EDGING

There is probably nothing so appropriate to the landscape of a garden as edging made of brick to separate sections. Note the effect in Fig. 13-30 and how one can vary the style to fit the effect desired. In the foreground is a thin edge of brick laid end-to-end. In the center the edging is brick laid face-to-face and on edge (rowlock style).

The job may be as difficult or as simple as desired, depending on how sturdy the construction is to be. The one described here is fairly easy to construct and will provide an edging that will stand

272

Fig. 13-29. Strike off excess top mortar, and follow up with a thorough cleanup job.

up under weathering for many years. The brick is to lie on a 1" thickness of concrete and the concrete on a 1" thickness of gravel for drainage. The same concrete base material will be used for mortaring the brick. A sand mix will do, which in this case was *Sakrete* sand mix. Mortar contains lime, which adds to its stiffness and allows building several courses of brick on top of each other. Since a brick edging is only one course, it is not necessary to use mixes containing lime. In fact, concrete mixes are stronger than mortar and will provide a strong bond and longer life.

Stake out a string line to act as a guide for digging a trench. Dig a trench the thickness of the brick, plus 1" for the concrete and 1" for the gravel. Lay a 1" layer of gravel along the bottom of the trench. Readjust the string line near the level of the ground and adjust it for a perfect level, using a carpenter's spirit level. Mix a quantity of premixed sand mix with water and lay a 1" bed over the gravel. Make enough to butter the joints between bricks.

Butter each brick with the concrete and lay them into place one

Fig. 13-30. Brick edging is excellent for separating garden sections.

at a time. Tap them into position, using the taut string as a guide. Allow about ⅜″ to ½″ between bricks for the joints. Trowel off excess concrete 15 minutes after laying to make the top smooth. You can add a course or two of brick above the ground level, for terracing or other effects. but if you do, the brick should be laid and jointed with mortar rather than concrete.

A BARBECUE GRILL

A barbecue grill which has many uses and is sturdily built is illustrated and fully dimensioned in Fig. 13-31. The top row of steel bars supports cooking utensils or steaks placed directly on the bars. The lower row of bars is for firewood and allows ashes to drop to the bottom. A brick ledge accommodates a slide-in pan for holding charcoal for this method of outdoor cooking.

A heavy concrete base extends 10″ into the ground, which eliminates any possibility of settling or of heaving in severe cold climates. Half-inch bars imbedded into the top section of the base will add to the strength to prevent cracking. Brick courses above the base follow the American bond pattern and, if well-mortared, will give years of service.

The concrete base is a 1:2:3½ mix; that is, the material proportions are 1 part portland cement, 2 parts graded sand, and 3½ parts graded aggregate. The masonry mortar is a 1:0.25:3 mix, or 1 part portland cement, ¼ part hydrated lime, and 3 parts sand. Because of the total amount of materials needed, it will be more economical to purchase them in bulk.

Here are the total materials for this particular barbecue grill:

Portland cement	3 sacks
Graded sand	9 cu. ft
Graded aggregate or gravel	6 cu. ft.
Hydrated lime	1 sack
Brick	250
3-ft. lengths, ½″ bars	4
2-ft. lengths, ½″ bars	26

Two sacks of cement will be used for the concrete and one sack for the mortar. If the ground is firm enough, no forms will be needed for the concrete base below ground level. A form 4¾ × 37½ (inside dimensions) should be made for that part of the base above ground to ensure smooth sides.

Dig out a rectangular hole to the dimensions mentioned above, leaving a 13½″ mound of dirt in the center. This mound saves on the amount of concrete needed. Mix the total amount of concrete in a power mixer, adding water as needed. About 17 gallons of water will be needed, and this should be precisely measured. Pour the concrete into the hole up to the ground level. Lay the 3-ft. rods in place at 12″ intervals. Pour the balance of the concrete up to the top of the forms in place. This should bring the base approximately 2″ above ground level. Carefully smooth and trowel the surface for a flat even top. Allow 58 hours for thorough setting of the concrete before starting the brickwork.

Carefully review the sketches in Fig. 13-31, and note that the fourth and sixth courses of brick have their inside tier slightly protruding inside the pit to support the lower row of rods and provide a ledge for the charcoal pan. Outside of this, the laying

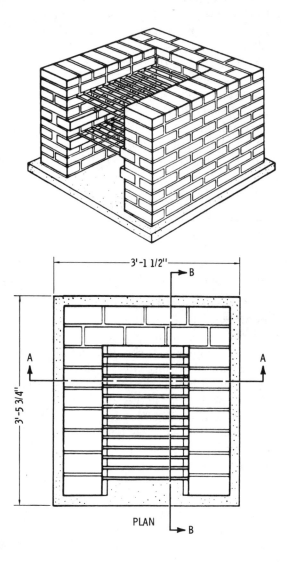

Fig. 13-31. Dimensions of a barbecue pit

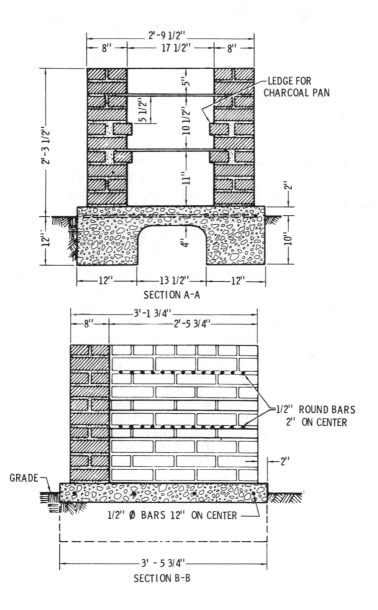

SECTION A-A

SECTION B-B

of brick and with a heavy concrete base.

277

up of the brick is standard, using ½" mortar joints. The outside edges of the brick structure should pass just 2" inside the edges of the concrete base. To make sure of this, it may be wise to dry-lay the first course of brick on all sides. Then pencilmark the outer edges for a guide. Another way is to drive stakes into the ground and use string guides on all four sides.

Mix an amount of mortar that can be handled within a 2-hour work session. Lay a bed of mortar on the concrete base for the first course of brick, using the pencilmarks or string guide to ensure straight sides. Because of the short wall sizes, it may not be necessary to use the taut line method described in previous chapters, but a straight board should be at hand to check the level of each course as it goes up.

Butter about five bricks at a time and put them in place, tapping them into position. After the fourth course is laid, place the 2-ft. rods in place over the protruding ledge of brick and space them parallel 2" apart. Carefully place mortar around them to hold them in place as the brick courses continue on up. Do not overlook the protruding bricks at the sixth course. Above the eighth course, place the second set of rods and mortar around them. Complete the brickwork, with the top two sides a course of all headers.

Be sure to trowel off excess mortar within 15 minutes after laying to prevent its setting up too hard to remove. Tool all joints firmly and clean off any feathers. To protect the base from the fallen mortar, place plastic sheets around the edges of the brick at the baseline to catch the debris. Otherwise, scrape off fallen mortar within 25 minutes after it has fallen onto the concrete base.

CHAPTER 14

Plasterboard and Plaster

Today, about 80% of all interior walls are surfaced with plasterboard, also known as Sheetrock (a brand name) and drywall; but plaster is still used, and there is no doubt that it makes a superior wall.

What follows is a consideration of plaster and plasterboard. One may never get to apply plaster, but it is well to know how it is applied as part of the general know-how of wall construction.

Two conventional methods are used to apply plaster to a wall. The older method is called *lath and plaster*, in which plaster is applied to properly spaced horizontally mounted furring strips. When applied in a thick coat, the plaster is forced into the spaces between the furring strips, forming a bond. The furring strips have been replaced, to a large extent, by metal screen lathing similar to that used for stucco or over gypsum lath.

Metal Lath

The sheets, usually measuring 27" × 96", have been die cut with numerous small slits and then "expanded," or pulled apart, to form openings for the keying of the plaster. The sheets come prepainted or galvanized to prevent rust and thus ensure a tight plaster bond.

Expanded metal lath can be used by itself as a plaster base in high-moisture applications or on furring strips installed over masonry suitable for direct plaster application. It can also be used to reinforce gypsum lath over window and door openings when cut into strips about 8" × 20" and lightly nailed in place.

Inside corners at the juncture of walls and ceilings should be reinforced with corners of metal lath or wire fabric, except where special clip systems are used for installing the lath. The minimum width of the lath in the corners should be 5", or 2½" on each surface or internal angle, and should also be lightly nailed in place. Corner beads of expanded metal lath or perforated metal should be installed on all exterior corners. They should be applied plumb and level. The bead acts as a leveling edge when the walls are plastered and reinforces the corner against mechanical damage.

Metal lath should be used under large flush beams and should extend well beyond the edges. Where reinforcing is required over solid wood surfaces, such as drop beams, the metal lath should either be installed on strips, or else self-furring nails should be used to set the lath out from the beam. The lath should be lapped on all adjoining gypsum lath surfaces.

Gypsum-board lath is generally 16" × 48" and is applied horizontally to the frame members. This type of board has a proper face with a gypsum filler. For studs and joists with a spacing of 16" on center, ⅜" thickness is used; and for 24" on center spacing, ½" thickness is used. This material can be obtained with a foil backing that serves as a vapor barrier, and if it faces an air space, it has some insulating value. It is also available with perforations, which improves the bonding strength of the plaster base.

Insulating fiberboard lath may also be used as a plaster base. It is usually ½" thick and generally comes in strips of 18" × 48". It often has a shiplap edge and may be used with metal clips that are located between studs and joists to stiffen the horizontal joints. Fiberboard lath has a value as an insulation and may be used on the walls or ceilings adjoining exterior or unheated areas.

Gypsum lath should be applied horizontally, with the joints broken, as shown in Fig. 14-1. Vertical joints should be made over the center of the studs or joists and should be nailed with 13-gauge gypsum lathing nails 1⅛″ long and having a ⅜″ flat head. Nails should be spaced 4″ on center and should be nailed at each stud or joist crossing. Lath joints over the heads of door and

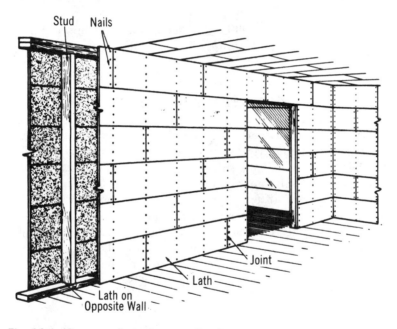

Fig. 14-1. Gypsum-plaster-base application.

window openings should not occur at the jamb lines. Insulating lath should be installed much like gypsum lath, except that 18-gauge, 1¼″ blued nails should be used.

The lath and plaster method requires a tremendous amount of labor and considerable time because three coats are applied and one must allow for drying time between coats. Since World War

II, this method has been almost entirely replaced by the plaster-board method, in which 4′ × 8′, 10′, or 12′ plasterboard sheets are nailed or screwed to the wall studs. Plasterboard (commonly called Sheetrock, a brand name) has a number of advantages over the lath and plaster method. It eliminates the need to mix plaster on the job site, which saves considerably on the amount of labor involved. Moreover, plasterboard is available in a large number of designs to fit almost any special requirement of interior wall finishing. It is made in thicknesses from ¼″ to ¾″. There are many suppliers of plasterboard throughout the country, and some make prefinished plasterboard to simulate various finishes.

Most of this chapter will be devoted to the installation of plasterboard. The lath and plaster method will be described only briefly.

PLASTER CONTENT

Plaster is generally cement, pure white in color, and includes a number of minerals. Most plasters contain the mineral gypsum, which is hydrated calcium sulfate. Some plasters with a smoother surface for the finish coat contain lime, which is calcium carbonate. Fine sand is often used in plaster, but not for the finish layer. Other minerals are included in smaller quantities, which alter the characteristics slightly, to meet certain requirements. For lath and plaster methods, plaster is supplied in bags as a chalk-like white powder. When mixed with water, it is plastic and workable. When the water evaporates, what remains is a rock-like hard plaster.

Plasterboards include a paper backing which helps to give the boards support for shipping and handling. This paper backing also gives a smooth finish for paint or wall paper in drywall construction.

LATH AND PLASTER

If you were to cut into the walls of older homes, you would find wood furring strips supporting a thick coat of plaster. A cutaway view is shown in Fig. 14-2. The wood furring strips are 4' long to fit studs 16" on center, 1 ½" wide, and ¼" thick. They are installed with a ⅜" space between each lath.

Wood furring strips have been replaced by metal lathing in the limited use today of the lath and plaster method. Metal lathing is die cut from flat stock sheets and then pulled apart to form a pattern like that shown in Fig. 14-3.

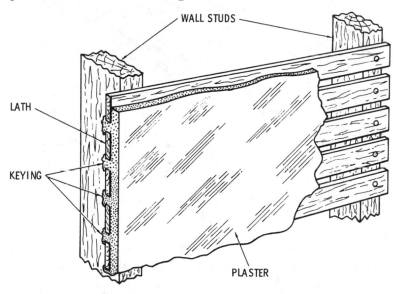

Fig. 14-2. Lath and plaster wall construction. Wood furring strips are nailed to the wall studs, and plaster is applied in several coats.

Plaster is supplied in bags with the necessary ingredients already mixed. It requires only the addition of water, which is added on the job. The amount of water required is marked on the bag. Plaster must be mixed in clean wooden boxes. Mix only as much as can be applied within an hour, or hydration will begin to take place and the plaster will become too hard to handle with ease. Each time a new batch is to be mixed, the box must first be thoroughly washed out. Leave the box clean at the end of the day ready for use the next day.

Plaster is carried to the work on a hand-held, square wooden platen and applied to the lathing with a rectangular trowel. It

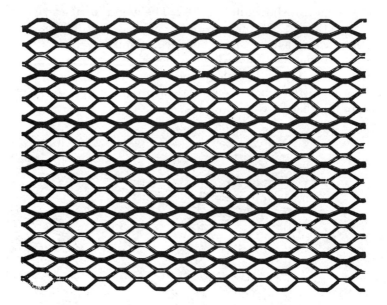

Fig. 14-3. Expanded metal screening like this is applied to the wall instead of wood furring strips.

must be firmly embedded into the lathing so that there is a firm keying into and around the edges of the lath strips.

Plasters are usually applied in three coats, as mentioned, which are termed in order of application as follows:

1. Scratch coat.
2. Second or brown coat.
3. Finish, or set, coat.

Scratch Coat—This coat should be made approximately ⅜″ thick, measured from the face of the backing, and carried to the full length of the wall or the natural breaking points, such as doors or windows. Before the scratch coat begins to harden, however, it should be cross-scratched to provide a mechanical key for the second or brown coat.

Brown Coat—The brown coat should be approximately the same thickness as the scratch coat. Before applying the brown coat, dampen the surface of the scratch coat evenly by means of a fog spray to obtain uniform suction. This coat may be applied in two thin coats, one immediately following the other. Such a method may prove helpful in applying sufficient pressure to ensure a proper bond with the base coat.

Bring the brown coat to a true, even surface, and then roughen with a wood flat or cross-scratch it lightly to provide a bond for the finish coat. Damp-cure the brown coat for at least two days, and then allow it to dry.

Finish Coat—The brown coat (as mentioned above) should be dampened for at least two days before application of the finish coat. Begin moistening as soon as the brown coat has hardened sufficiently, applying the water in a fine fog spray. Avoid soaking the wall, but give it as much water as will readily be absorbed.

Applying the Plaster

The assortment of tools used by the plasterer are very similar to those employed by the bricklayer. Essential plastering tools are as follows:

Hawk—This tool is usually made of hard pine or cedar and is usually about 13″ or 14″ square (Fig. 14-4). It is held in the left hand, forming a small "hand table" which holds a supply of plaster. The plasterer carries this plaster from the hawk to the work, where he spreads it over the surface to be plastered, as shown in Fig. 14-4.

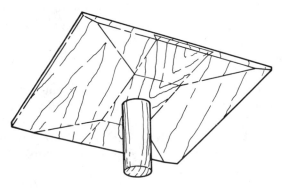

Fig. 14-4. A wood hawk used to carry the plaster material to the wall or ceiling.

Trowel—As distinguished from the bricklayer's trowel, the plasterer's trowel is rectangular, as shown in Fig. 14-5. The handle is attached to a mounting which stiffens the blade. The trowel is light, so that the tool is easily used. Trowels are classed as:

1. Browning.
2. Finishing.

The browning trowel is used for rough coating and has a heavier blade than that used on the finishing trowel; otherwise the construction of both are the same.

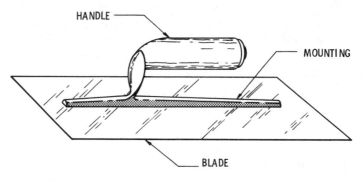

Fig. 14-5. The trowel which is used to apply plaster is generally made of 24-gauge polished steel.

Float—The common form of float consists of a piece of hard pine board 10″ or 12″ × ⅝″ to ⅞″, having a wooden handle, preferably of hardwood screwed to the back. Because of the great friction, the face of a float soon wears off and becomes thin; hence, there is usually an adjustable handle fastened with bolts which can be fixed to new face pieces as they are required. Floats are applied in smoothing and finishing with a rotary motion, sometimes reversed as left to right, and vice versa. Various types of floats are shown in Fig. 14-6.

Darby—This tool is simply a flat straight strip of wood (or metal) provided with handles to enable the workman to level up and straighten large surfaces as they are put on.

The tool is held by both hands and moved with a sliding up-and-down diagonal and horizontal motion to level off by rubbing and pressing any lumps or high spots which may be left after

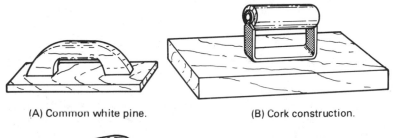

(A) Common white pine.

(B) Cork construction.

(C) Cork-faced float with hard wood backing.

(D) Wood angle.

Fig. 14-6. Several types of wooden floats.

applying the mortar with trowels. This work is very laborious, especially on ceilings or any job above the line of the shoulders. The tool is also essential in preserving an even thickness of each coat. Fig. 14-7 shows an ordinary wooden darby.

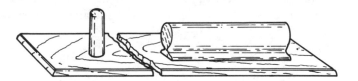

Fig. 14-7. An ordinary wooden darby used to level large areas of plaster.

It is important that the first coat of plaster have the right consistency to establish a firm key onto the lath. If the plaster is too thick, it will stand out behind the lathing material but not grab securely. A thinner plaster tends to turn down behind the lathing material and form a key and lock into place. This is illustrated in Fig. 14-8 with wood lathing. The same reasoning applies to metal lathing, although keying to expanded metal lathing is more secure even with stiff plaster material.

287

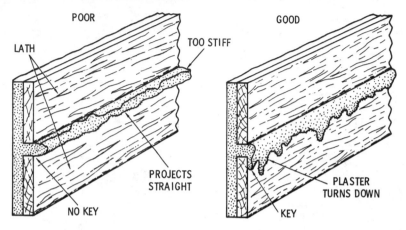

Fig. 14-8. Good and bad embedding of plaster on furring strips. The plaster must form a key behind the strips to hold well.

Each coat must be throughly dry before applying the next coat. While the second coat is fogged with a spray of water before applying the finish coat, the second coat must be allowed to dry before fogging.

It is often desirable to obtain a hard, glossy surface on the finish coat. This is done by brushing water on the surface, using a large painter's brush, and then floating again with a hardwood float. If this is done a couple of times, the surface will take on a very hard finish.

One-Coat Plastering

In addition to the plastering methods described, there has been one other development in plastering that reduces the plastering process to "one coat" and that gives the process its name. Here, "green" or "blue" ⅝" gypsum board is installed like regular gypsum board, and the premixed plaster material, which comes in 5-gal. cans, is applied by the plasterer. That's it! The applied material cures very hard, and the gypsum board has the tooth to keep it in place.

While the description given above for applying plaster is brief, it can readily be seen how much labor is involved. This is why this kind of plastering has given way so often to the use of plasterboard.

PLASTERBOARD

Plasterboard [also referred to as Sheetrock (a brand name), drywall, and wallboard] is available in solid sheets made of fireproof material. Standard width is usually 4 ft., with a few types in 2-ft. widths. Lengths are from 6 to 16 ft., and thicknesses are from $\frac{1}{4}$" to $\frac{3}{4}$". Edges are tapered to permit smooth edge finishing, although some boards can be purchased with square edges.

The most common length is 8 ft. to permit wasteless mountings to the wall studs. This particular size will fit standard 8-ft.-ceiling heights when vertically mounted. However, recommended mounting for easier handling is horizontal. Two lengths, one above the other, fit 8-ft.-ceiling heights.

Regular wallboard comes in thicknesses of $\frac{1}{4}$", $\frac{3}{8}$", $\frac{1}{2}$", and $\frac{5}{8}$". The $\frac{1}{4}$" board is used in two- or three-layer wallboard construction and generally has square edges. The most commonly used thickness is $\frac{3}{8}$", which can be obtained in lengths up to 16 ft. All have a smooth, cream-colored, paper covering that takes any kind of decoration.

Insulating (foil-backed) wallboards have aluminium foil which is laminated to the back surface. The aluminum foil creates a vapor barrier and provides reflected insulation value. It can be used in a single-layer construction or as a base layer in two-layer construction. Thicknesses are $\frac{3}{8}$", $\frac{1}{2}$", and $\frac{5}{8}$", with lengths from 6 to 14 ft. All three thicknesses are available with either square or tapered edges.

Some suppliers have lower-cost boards called *backer board.* These are used for the base layer in two-layer construction. Thicknesses are $\frac{3}{8}$", $\frac{1}{2}$", and $\frac{5}{8}$", with a width of 2 ft. as well as the usual 4 ft. Length is only 8 ft. The edges are square, but the 2-ft.-wide board, in $\frac{1}{2}$" and $\frac{5}{8}$" thicknesses, can be purchased with tongue-and-groove edges.

Another type of backer board has a vinyl surface. It is used as a waterproof base for bath and shower areas. The size is 4 ft. × 11 ft., which is the right size for enclosing around a standard bathtub. It is available in ½″ and ⅝″ thicknesses with square edges.

A moisture-resistant board is especially processed for use as a ceramic tile base. Both core and facing paper are treated to resist moisture and high humidity. Edges are tapered, and the regular tile adhesive seals the edges. The boards are 4 ft. wide, with lengths of 8 and 12 ft.

There is also vinyl-covered plasterboard, which eliminates the need for painting or wallpapering. The vinyl covering is available in a large number of colors and patterns to fit many decorator needs. Many colors in a fabric-like finish plus a number of wood-grain appearances can be obtained. Only mild soap and water are needed to keep the finish clean and bright. Usual installation is by means of an adhesive, which eliminates nails or screws except at the top and bottom where decorator nails are generally used. The square edges are butted together (Fig. 14-9). They are available in ½″ and ⅜″ thicknesses, 4 ft. wide, and 8, 9, and 10 ft. long.

Fig. 14-9. Home interior where vinyl-finish plasterboard is used.

A similar vinyl-covered plasterboard, called *monolithic*, has an extra width of vinyl along the edges. The boards are installed with adhesive or nails. The edges are brought over the fasteners, pasted down, and then cut flush at the edges. This method hides the fasteners. Extruded aluminum bead and trim accessories, covered with matching vinyl, can be purchased to finish off inside, outside, and ceiling corners, for vinyl-covered plasterboard.

Plasterboard Ratings

The principal manufacturers of plasterboard maintain large laboratories for the testing of their products for both sound-absorption ability and fire retardation. They are based on established national standards that have been industry-accepted, and all follow the same procedure.

CONSTRUCTION WITH PLASTERBOARD

Several alternative methods of plasterboard construction may be used, depending on the application and desires of the customer or on yourself, if you are the homeowner. Most residential homes use $2'' \times 4''$ wall studs, and single-layer plasterboard walls are easily installed. Custom-built homes, commercial buildings, and party walls between apartments should use double-layer plasterboard construction. It provides better sound insulation and fire retardation. For the greatest protection against fire hazards, steel frame partitions plus plasterboard provide an all-noncombustible system of wall construction. All-wood or the all-steel construction applies to both walls and ceilings.

Plasterboard may also be applied directly to either insulated or uninsulated masonry walls. Plasterboard may be fastened in place by any of several methods, such as nails, screws, or adhesives. In addition, there are special resilient furrings and spring clips, both providing added sound deadening.

Single Layer on Wood Studs

The simplest method of plasterboard construction is a single layer of plasterboard nailed directly to the wood studs and ceiling joists (Fig. 14-10). For single-layer construction, ⅜", ½", or ⅝" plasterboard is recommended.

The ceiling boards should be installed first, and then the wall boards. Install plasterboard perpendicular to the studs for minimum joint treatment and greater strength. This applies to the ceiling as well. Either nails, screws, or adhesive may be used to

Fig. 14-10. The most common interior finishing today uses plasterboard; it is known as *drywall construction.*

fasten the plasterboard. Blue wallboard nails have annular rings for better grip than standard nails, and the screws have Phillips heads for easy installation with a power driver. In single-nail installations, as shown in Fig. 14-11, the nails should be spaced not to exceed 7" apart on the ceilings and 8" on the walls. An alternative method, one which reduces nail popping, is the double-nail method. Install the boards in the single-nail method first, but with nails 12" apart. Then drive a second set of nails about 1½″ to 2″ from the first (Fig. 14-11), from the center out, but not at the perimeter. The first series of nails are then struck again to ensure the board being drawn up tight. Use 4d cooler type nails for ⅜″ regular foil backed or backer board if used, 5d for ½″ board, and 6d for ⅝″ board.

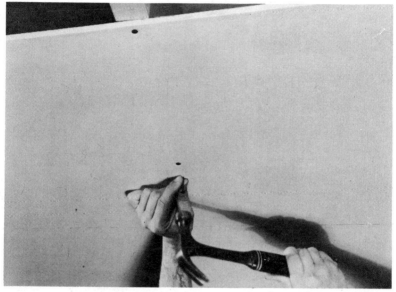

Courtesy National Gypsum Co.

Fig. 14-11. In the double-nail method of fastening wallboard, the first set of nails are put in place and a second set of nails are driven in place about 1½″ to 2″ from the first.

Screws are the preferred method of fastening plasterboard since they push the board up tight against the studding and will not loosen. Where studs and joists are 16″ on center, the screws can be 12″ apart on ceilings and 16″ apart on walls. If the studs are 24″ on center, the screws should be not over 12″ apart on the walls. Fig. 14-12 shows a power screwdriver in use. Strike the heads of all nails and screws to just below the surface of the board. The dimples will be filled in when joints and corners are finished.

A number of adhesives are available for installing plaster-

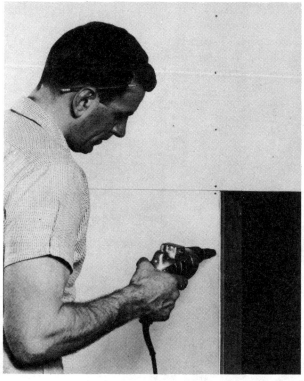

Courtesy National Gypsum Co.

Fig. 14-12. Wallboard fastened to studs with a power screwdriver.

board. Some are quick-drying; others are slower. Adhesives can be purchased in cartridge or bulk form, as shown in Fig. 14-13. The adhesive is applied in a serpentine bead, as shown in Fig. 14-14, to the facing edges of the studs and joists. Place the wall board in position, and nail it temporarily in place. Use double-headed nails or nails through a piece of scrap plasterboard. When the adhesive has dried, the nails can be removed and the holes filled when sealing the joints.

Plasterboard is butted together but not forced into a tight fit. The treatment of joints and corners is covered later. The adhesive method is ideal for prefinished wallboard, although special nails with colored heads can be purchased for the decorator boards.

Fig. 14-13. Adhesives are available either in bulk or in cartridge form.

Do not apply more adhesive than can permit installation within 30 minutes.

Two-Layer Construction

As mentioned, two layers of plasterboard improve sound deadening and fire retardation. The first layer of drywall or plasterboard is applied as described before. Plasterboard may be less expensive used as a backing-type board. Foil-backed or special sound-reducing board may also be used. Nails need only be struck and screws driven with their heads just flush with the surface of the board. The joints will not be given special treatment since they will be covered by the second layer.

The second layer is cemented to the base layer. The facing layer is placed over the base layer and temporarily nailed at the top and bottom. Temporary nailing can be done with either double-headed nails or blocks of scrap wood or plasterboard

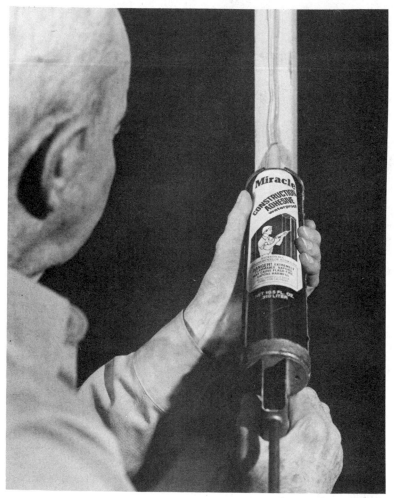

Fig. 14-14. Using a cartridge to apply a bead of adhesive to the wall studs.

under the heads. The face layer is left in place until the adhesive is dry.

To make sure of good adhesion, lay bracing boards diagonally from the center of the facing layer to the floor. Prebowing of the boards is another method. Lay the plasterboard finish face down across a 2″ × 4″ for a day or two. Let the ends hang free.

Joints of boards on each layer should not coincide but should be separated by about 10″. One of the best ways of doing this is to install the base layer horizontally and the finish layer vertically. When using adhesive, a uniform temperature must be maintained. If construction is in the winter, the rooms should be heated to somewhere between 55 and 70°F and kept well ventilated.

Double-layer construction, with adhesive, is the method used with decorator-type, vinyl-finished paneling, except that special matching nails are used at the base and ceiling lines and left permanently in place.

Fig. 14-15 shows the basic double-layer construction, with details for handling corners and an alternative method of wall construction for improved sound deadening.

Other Methods of Installation

Sound transmission through a wall can be further reduced if the vibrations of the plasterboard can be isolated from the studs or joists. Two methods are available to accomplish this. One uses metal furring strips with edges formed in such a way as to give them flexibility. The other is by means of metal push-on clips with bowed edges also for flexibility, used on ceilings only.

Fig. 14-16 shows the appearance and method of installing resilient furring channels. They are 12-ft.-long strips of galvanized steel and include predrilled holes spaced every inch. This permits nailing to studs 16″ or 24″ on center. Phillips-head, self-tapping screws are power-driven through the plasterboard and into the surface of the furring strips.

As shown in Fig. 14-16, strips of 3″ × ½″ plasterboard are fastened to the sole and plate at top and bottom to give the plasterboard a solid base. For best sound isolation, the point of intersection between the wall and floor should be caulked prior to application of the baseboard.

On ceiling joists, use two screws at each joist point to fasten the furring strips. Do not overlap ends of furring strips; leave about a ⅜″ space between ends. Use only ½″ or ⅝″ wallboard with this system.

Suspending plasterboard ceilings from spring clips is a method

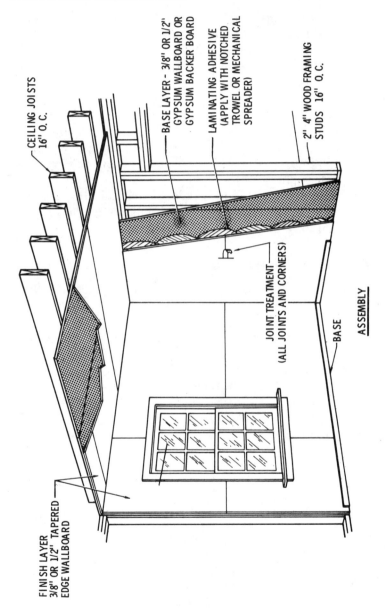

CEILING JOISTS 16" O. C.

BASE LAYER – 3/8" OR 1/2" GYPSUM WALLBOARD OR GYPSUM BACKER BOARD

LAMINATING ADHESIVE (APPLY WITH NOTCHED TROWEL OR MECHANICAL SPREADER)

2" 4" WOOD FRAMING STUDS 16" O. C.

JOINT TREATMENT (ALL JOINTS AND CORNERS)

BASE

ASSEMBLY

FINISH LAYER 3/8" OR 1/2" TAPERED EDGE WALLBOARD

Fig. 14-15. A cross-sectional view of double-layer wall construction, details

298

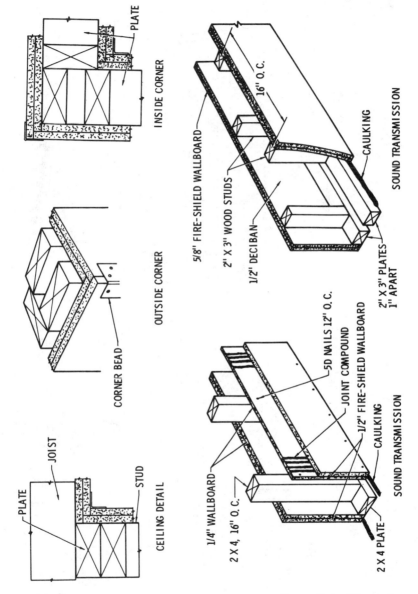

PLATE

INSIDE CORNER

OUTSIDE CORNER

CORNER BEAD

JOIST

PLATE

STUD

CEILING DETAIL

5/8" FIRE-SHIELD WALLBOARD

2" X 3" WOOD STUDS

1/2" DECIBAN

16" O. C.

CAULKING

SOUND TRANSMISSION

2" X 3" PLATES
1" APART

5D NAILS 12" O. C.

JOINT COMPOUND

1/2" FIRE-SHIELD WALLBOARD

CAULKING

SOUND TRANSMISSION

1/4" WALLBOARD

2 X 4, 16" O. C.

2 X 4 PLATE

Courtesy National Gypsum Co.

for handling corners, and special wall construction for sound deadening.

299

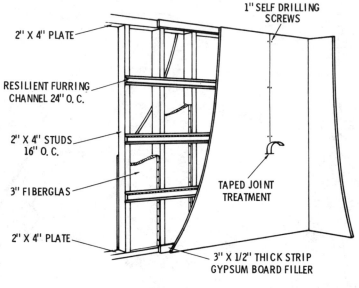

PERSPECTIVE

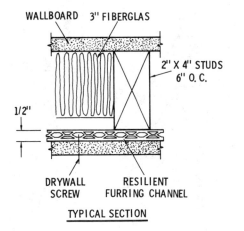

TYPICAL SECTION

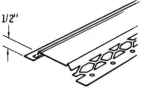

RESILIENT FURRING
CHANNEL

Fig. 14-16. Resilient furring strips, with expanded edges, may be installed to isolate the plasterboard from the studs.

of isolating the vibrations from the ceiling joists and floor above. Details are shown in Fig. 14-17.

The push-on clips are placed on 1″ × 2″ nominal size wood-furring strips. The clips are then nailed to the sides of the ceiling joists with short annular ringed or cooler type nails. With furring strips installed, they will be against the nailing edge of the joists. This provides a solid foundation for nailing the plasterboard to the clips. The weight of the plasterboard after installation will stretch the clips to a spring position.

The wood furring strips must be of high-quality material so that they will not buckle, twist, or warp. Nails for attaching the wallboard should be short enough so they will not go through the furring strips and into the joist edges. The wallboard must be held firmly against the strips while it is being nailed. Expansion joints must be provided every 60 ft. or for every 2400 sq. ft. of surface.

It is essential that the right number of clips be used on a ceiling, depending on the thickness of the wallboard and whether it is single- or double- layered. The spring or efficiency of the clips is affected by the weight.

Tile Underlayment

At least two types of plasterboard are available as backing material for use in tiled areas where moisture protection is important, such as tub enclosures, shower stalls, powder rooms, kitchen-sink splash boards, and lockerrooms. One is a vinyl-covered plasterboard, and the other is a specially designed board material to prevent moisture penetration.

The vinyl-covered board is 4 ft. × 11 ft., the right length for a tube enclosure. By scoring it and snapping it to length, the vinyl covering can be a continuous covering all around the tub, and no corner sealing is required. This is shown in one of the sketches in Fig. 14-18. This board has square edges and is nailed or screwed to the studs without the need to countersink the nails or screws. The waterproof tile adhesive is applied, and the tile is installed over the board. This board is intended for full tile treatment and should not extend beyond the edges of the tile.

Unlike vinyl-surfaced board, the moisture-resistant board may

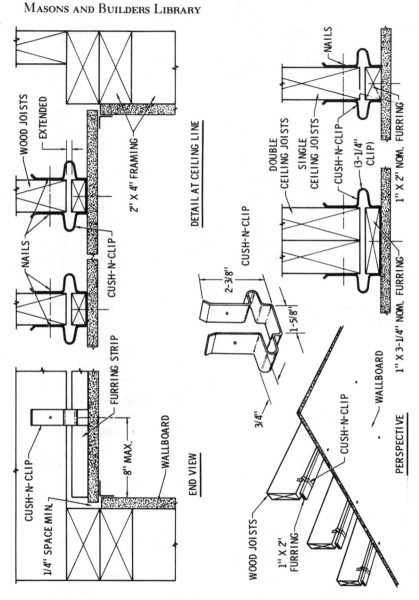

Courtesy National Gypsum Co.

Fig. 14-17. Details for installing spring clips to ceiling joist. Clips suspend plasterboard from joist to provide isolation from noise vibrations.

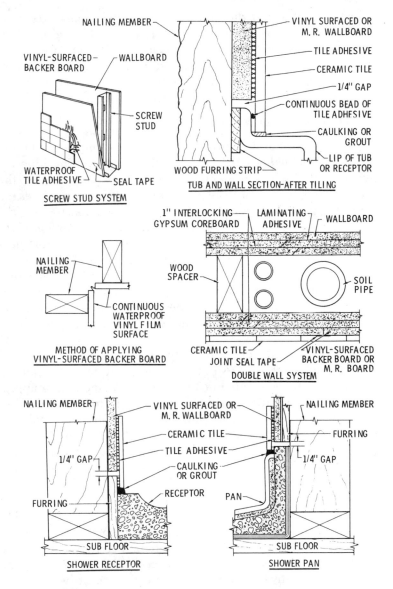

NAILING MEMBER

VINYL-SURFACED BACKER BOARD

WALLBOARD

SCREW STUD

WATERPROOF TILE ADHESIVE

SEAL TAPE

SCREW STUD SYSTEM

VINYL SURFACED OR M. R. WALLBOARD

TILE ADHESIVE

CERAMIC TILE

1/4" GAP

CONTINUOUS BEAD OF TILE ADHESIVE

CAULKING OR GROUT

LIP OF TUB OR RECEPTOR

WOOD FURRING STRIP

TUB AND WALL SECTION-AFTER TILING

NAILING MEMBER

CONTINUOUS WATERPROOF VINYL FILM SURFACE

METHOD OF APPLYING VINYL-SURFACED BACKER BOARD

1" INTERLOCKING GYPSUM COREBOARD

LAMINATING ADHESIVE

WALLBOARD

WOOD SPACER

SOIL PIPE

CERAMIC TILE

JOINT SEAL TAPE

VINYL-SURFACED BACKER BOARD OR M. R. BOARD

DOUBLE WALL SYSTEM

NAILING MEMBER

VINYL SURFACED OR M. R. WALLBOARD

CERAMIC TILE

TILE ADHESIVE

CAULKING OR GROUT

1/4" GAP

RECEPTOR

FURRING

SUB FLOOR

SHOWER RECEPTOR

NAILING MEMBER

FURRING

1/4" GAP

PAN

SUB FLOOR

SHOWER PAN

Fig. 14-18. Details for installing vinyl-locked and moisture-resistant wallboard for bath and shower use.

303

be extended beyond the area of the tile. The part extending beyond the tile may be painted with latex or oil-based paint, or papered. It has tapered edges and is installed and treated in the same manner as regular plasterboard. Corners are made waterproof by the tile adhesive. The sketches in Fig. 14-18 show details for the installation of both types of plasterboard.

Plasterboard over Masonry

Plasterboard may be applied to inside masonry walls in any one of several methods: directly over the masonry using adhesive cement; over wood or metal furring strips fastened to the masonry; to polystyrene insulated walls by furring strips over the insulation; or by lamination directly to the insulation.

Regular or prefinished plasterboard may be laminated to unpainted masonry walls, such as concrete or block interior partitions, and to the interior of exterior walls above or below grade. While nearly all of the adhesive available may be used either above or below grade, the regular joint compounds are recommended for use only above grade. Exterior masonry must be waterproofed below grade and made impervious to water above grade.

The masonry must be clean and free of dust, dirt, grease, oil, loose particles, or water-soluble particles. It must be plumb, straight, and in one plane. Fig 14-19 shows the method of applying the adhesive to masonry walls. Boards may be installed either horizontally or vertically. Ceiling wallboard should normally be applied last to allow nailing temporary bracing to the wood joists. The wallboard should be installed with a clearance of ⅛" or more from the floor to prevent wicking. If there are expansion joints in the masonry, cut the wallboard to include expansion joinsts to match that of the masonry.

The sketches of Fig. 14-20 show the installation of wallboard to masonry with furring strips. Furring strips may be wood or U-shaped metal as used for foam insulation. The furring strips are fastened to the masonry with concrete nails. They may be mounted vertically or horizontally, but there must be one horizontal strip along the base line. The long dimensions of the wallboards are to be perpendicular to the furring strips. Furring strips are fastened 24" on center. The wallboard is attached to the

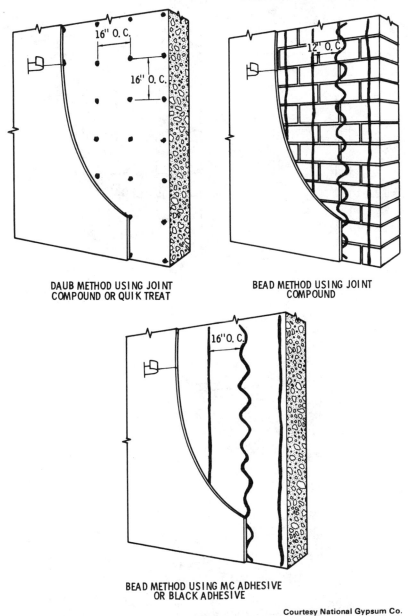

DAUB METHOD USING JOINT
COMPOUND OR QUIK TREAT

BEAD METHOD USING JOINT
COMPOUND

BEAD METHOD USING MC ADHESIVE
OR BLACK ADHESIVE

Courtesy National Gypsum Co.

Fig. 14-19. Method of applying adhesive directly to masonry walls.

305

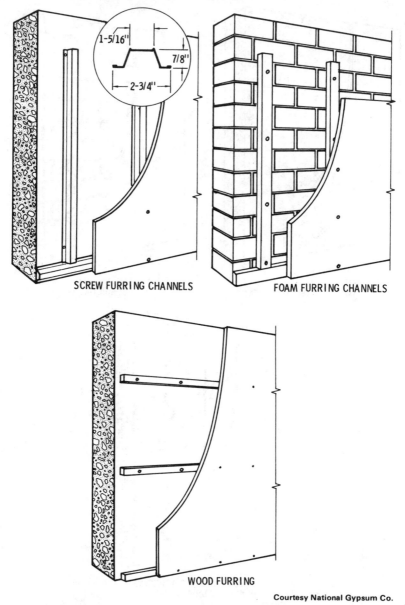

SCREW FURRING CHANNELS

FOAM FURRING CHANNELS

WOOD FURRING

Courtesy National Gypsum Co.

Fig. 14-20. Three types of furring strips that are used between masonry walls and plasterboard.

306

furring strips in the usual way. Use self-drilling screws on the metal furring strips and nails or screws on the wood strips.

Insulation may be applied between the plasterboard and the masonry, using urethane foam sheets or extruded polystyrene. Plasterboard is then secured to the wall either with the use of furring strips over the insulation or by laminating directly to the insulation.

Special U-shaped metal furring strips are installed over the insulation, as shown in Fig. 14-21. The insulation pads may be

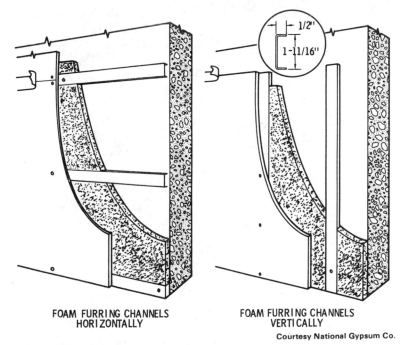

FOAM FURRING CHANNELS
HORIZONTALLY

FOAM FURRING CHANNELS
VERTICALLY

Courtesy National Gypsum Co.

Fig. 14-21. U-shaped furring channels installed horizontally or vertically over foam-insulating material.

held against the wall while the strips are installed, or they can be held in place with dabs of plasterboard adhesive. Use a fast-drying adhesive. If some time is to elapse before the furring strips and plasterboard are to be installed, dot the insulation with adhesive every 24″ in both directions on the back of the insulation pads.

The metal furring strips are fastened in place with concrete nails. Place them a maximum of 24″ apart, starting about 1″ to 1½″ from the ends. The strips may be mounted vertically or horizontally, as shown in Fig. 14-21. Fasten the wallboard to the furring strips with a power driver (Fig. 14-22), using self-drilling screws. Screws must not be any longer than the thickness of the wallboard plus the ½″ furring strip. Space screws 24″ on center for ½″ and ⅝″ board, 16″ on center for ⅜″ board.

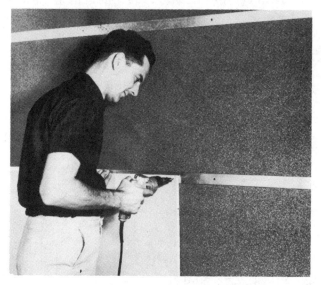

Fig. 14-22. Special self-drilling screws are power-driven through the plasterboard and into the metal furring strip.

Plasterboard may be laminated directly to the foam insulation. The sketches in Fig. 14-23 show how this is done for both horizontal and vertical mounted boards. Install wood furring strips onto the masonry wall for the perimeter of the wallboard. The strips should be 2″ wide and 1/32″ thicker than the foam insulation. Include furring where the boards join, as shown in the left-hand sketch of Fig. 14-23.

Apply a ⅜″ diameter bead of adhesive over the back of the foam insulation and in a continuous strip around the perimeter.

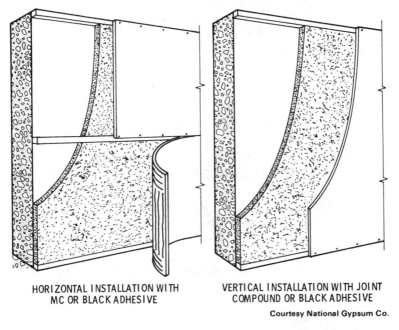

HORIZONTAL INSTALLATION WITH
MC OR BLACK ADHESIVE

VERTICAL INSTALLATION WITH JOINT
COMPOUND OR BLACK ADHESIVE

Courtesy National Gypsum Co.

Fig. 14-23. Plasterboard may be installed against the foam insulation with adhesive.

Put the bead dots about 16″ apart. Apply the foam panels with a sliding motion, and hand-press the entire panel to ensure full contact with the wall surface. For some adhesives, it is necessary to pull the panel off of the wall to allow flashoff of the solvent. Then reposition the panel. Read the instructions that come with the adhesive.

Coat the back of the wallboard and press against the foam insulation. Nail or screw the boards to the furring strips. The nails or screws must not be too long, or they will penetrate the furring strips and press against the masonry wall. The panels must clear the floor by ⅛″. Almost any type of adhesive may be used with urethane foam, but some adhesives cannot be used with polystyrene. Also, in applying prefinished panels, be careful about the choice of adhesive. Some of the quick-drying types are harmful to the finish. Read the instructions on the adhesive before you buy.

FIREPROOF WALLS AND CEILINGS

For industrial application, walls and ceilings may be made entirely of noncombustible materials using galvanized steel wall and ceiling construction and faced with plasterboard. Assembly is by self-drilling screws, driven by a power screwdriver with a Phillips bit.

WALL PARTITIONS

Construction consists of U-shaped track which may be fastened to existing ceilings, or the steel member ceilings, and to the floors. U-shaped steel studs are screwed to the tracks. Also available are one-piece or three-piece metal door frames. The three-piece frames are usually preferred by contractors since they permit finishing the entire wall before the door frames are installed.

Fig. 14-24 details wall construction as well as the cross section of the track and stud channels. For greater sound deadening, double-layer board construction with resilient furring strips between the two layers is recommended. This is shown in a cut-away view in the lower-right-hand corner of Fig. 14-24.

Figs. 14-25 and 14-26 show details for intersecting walls, jambs, and other finishing needs. Table 14-1 lists door-frame specifications. Chase walls are used (Fig. 14-27), where greater interior wall space is needed (between the walls) for equipment.

Fig. 14-28 shows how brackets are attached for heavy loads. Table 14-2 lists the allowable load for bolts installed directly to the plasterboard.

STEEL-FRAME CEILINGS

Three types of furring members are available for attaching plasterboard to ceilings. They are screw-furring channels, resilient screw-furring channels, and screw studs. Self-drilling screws attach the wallboard to the channels. These are illustrated in Fig.

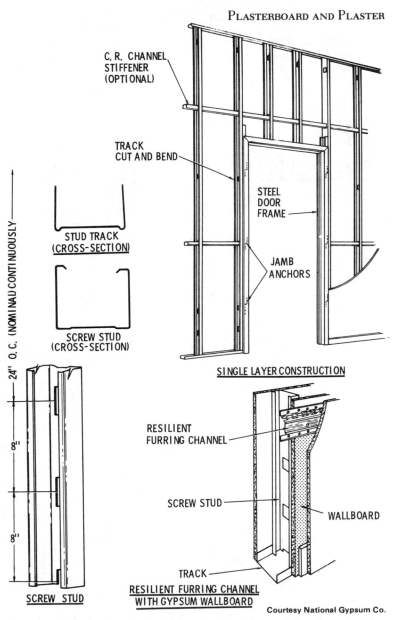

C. R. CHANNEL STIFFENER (OPTIONAL)

TRACK CUT AND BEND

STEEL DOOR FRAME

JAMB ANCHORS

STUD TRACK (CROSS-SECTION)

SCREW STUD (CROSS-SECTION)

24" O. C. (NOMINAL) CONTINUOUSLY

8"

8"

SCREW STUD

SINGLE LAYER CONSTRUCTION

RESILIENT FURRING CHANNEL

SCREW STUD

WALLBOARD

TRACK

RESILIENT FURRING CHANNEL WITH GYPSUM WALLBOARD

Courtesy National Gypsum Co.

Fig. 14-24. Details of metal wall construction using metal studs and tracks. This type of wall construction becomes an all-noncombustible interior wall.

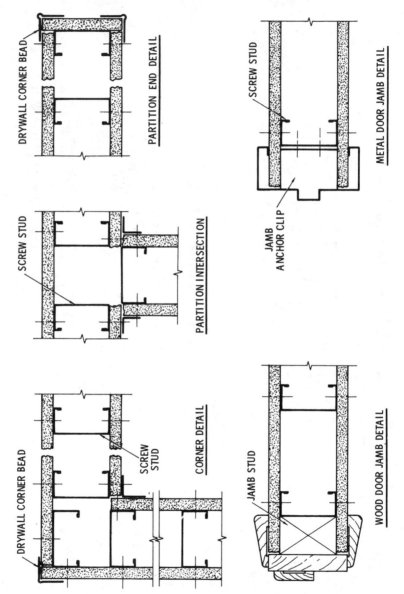

Courtesy National Gypsum Co.

Fig. 14-25. Details for intersections and jambs using metal wall construction.

14-29. Any of the three may be attached to the lower chord of steel joists or carrying channels in suspended ceiling construction. Either special clips or wire ties are used to fasten the channels. Fig. 14-30 shows the details in a complete ceiling assembly.

SEMISOLID PARTITIONS

Complete partition walls may be made of all plasterboard at a considerable savings in material and time over the regular wood and plasterboard types. These walls are thinner than the usual $2'' \times 4''$ stud wall, but they are not load-bearing walls.

Fig. 14-31 shows cross-section views of the all-plasterboard walls, with details for connecting them to the ceiling and floor. Also shown is a door frame and a section of the wall using the baseboard. Partitions may be $2\frac{1}{4}''$, $2\frac{5}{8}''$, or $2\frac{7}{8}''$ thick, depending on the thickness of the wallboard used and the number of layers. The centerpiece is not solid, but a piece of $1''$ or $1\frac{5}{8}''$ plasterboard about $6''$ wide acting as a stud element.

Fig. 14-32 shows how the wall is constructed. The boards are vertically mounted, and so their length must be the same distance as from the floor to the ceiling, but not exceeding 10 ft. The layers are prelaminated on the job, and then raised into position.

Place two pieces of wallboard on a flat surface with face surfaces facing each other. These must be the correct length for the height of the wall. Cut two studs $6''$ wide and a little shorter than the length of large wallboards. Spread adhesive along the entire length of the two plasterboard studs, and put one of them, adhesive face down, in the middle of the top large board. It should be $21''$ from the edges and equidistant from the ends. Set the other stud, adhesive face up, flush along one edge of the large plasterboard.

Place two more large boards on top of the studs, with the edge of the bundle in line with the outer edge of the uncoated stud. Temporarily place a piece of plasterboard under the opposite edge for support. In this system each pair of boards will not be directly over each other, but alternate bundles will protrude. Continue the procedure until the required number of assemblies

313

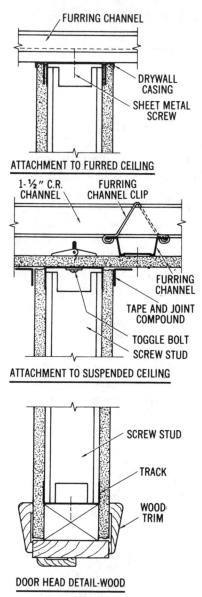

FURRING CHANNEL

DRYWALL CASING

SHEET METAL SCREW

ATTACHMENT TO FURRED CEILING

1-½" C.R. CHANNEL

FURRING CHANNEL CLIP

FURRING CHANNEL

TAPE AND JOINT COMPOUND

TOGGLE BOLT

SCREW STUD

ATTACHMENT TO SUSPENDED CEILING

SCREW STUD

TRACK

WOOD TRIM

DOOR HEAD DETAIL-WOOD

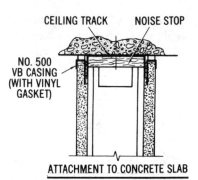

CEILING TRACK   NOISE STOP

NO. 500 VB CASING (WITH VINYL GASKET)

ATTACHMENT TO CONCRETE SLAB

8" MIN.

PAN HEAD SCREW

STUD SPLICE

Fig. 14-26. How to handle

is obtained. Let dry, with temporary support under the overhanging edges. Fig. 14-32 shows the lamination process.

To raise the wall, first install a 24"-wide starter section of wallboard with one edge plumbed against the intersecting wall. Spread adhesive along the full length of the plasterboard stud of one wall section, and erect it opposite to the 24" starter section. The free edge of the starter section should center on the stud

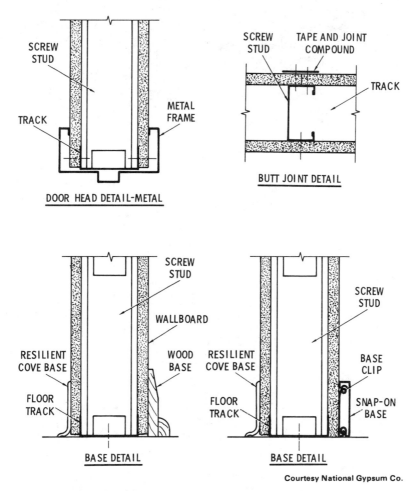

DOOR HEAD DETAIL-METAL

BUTT JOINT DETAIL

BASE DETAIL

BASE DETAIL

Courtesy National Gypsum Co.

ceiling and base finishing.

piece. Apply adhesive to the plasterboard stud of another section, and erect it adjacent to the starter panel. Continue to alternate sides as you put each section in place.

Some plasterboard manufacturers make thick, solid plasterboards for solid partition walls. Usually they will laminate two pieces of 1"-thick plasterboard to make a 2"-thick board. Chase

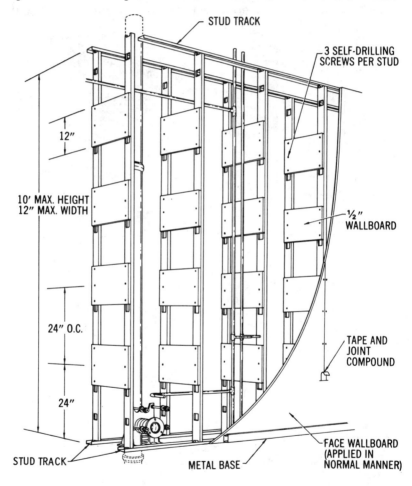

CHASE WALL

Fig. 14-27. Deep-wall construction where space is needed for various

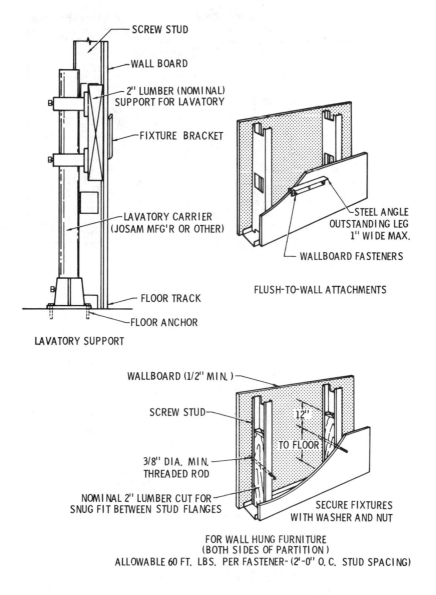

SCREW STUD

WALL BOARD

2" LUMBER (NOMINAL)
SUPPORT FOR LAVATORY

FIXTURE BRACKET

LAVATORY CARRIER
(JOSAM MFG'R OR OTHER)

FLOOR TRACK

FLOOR ANCHOR

LAVATORY SUPPORT

STEEL ANGLE
OUTSTANDING LEG
1" WIDE MAX.

WALLBOARD FASTENERS

FLUSH-TO-WALL ATTACHMENTS

WALLBOARD (1/2" MIN.)

SCREW STUD

12"
TO FLOOR

3/8" DIA. MIN.
THREADED ROD

NOMINAL 2" LUMBER CUT FOR
SNUG FIT BETWEEN STUD FLANGES

SECURE FIXTURES
WITH WASHER AND NUT

FOR WALL HUNG FURNITURE
(BOTH SIDES OF PARTITION)
ALLOWABLE 60 FT. LBS. PER FASTENER- (2'-0" O.C. STUD SPACING)

Courtesy National Gypsum Co.

reasons, such as plumbing and heating pipes.

Table 14-1. Door and Frame Selector (Type of Door Frame)

Door Weight	Gauge Jamb Studs	Bracing Over Header	Knock Down Alum.	Knock Down Steel	Fixed Steel	Requires Mechanical Closure	Height* Stud Size			
							1 5/8"	2 1/2"	3 5/8"	4"
Up to 50 lb.	25		x				NA†	10'	14'	15'
	25			x			NA†	10'	14'	15'
	25				x		NA†	10'	14'	15'
	25	x	x				10'	12'	16'	17'
	25	x		x			10'	12'	16'	17'
	25	x			x		10'	12'	16'	17'
	20	x	x				—	16'	21'	22'
	20	x		x			—	16'	21'	22'
	20	x			x		—	16'	21'	22'
50 to 80 lb.	25	x			x	x	12'	10'	14'	15'
	25	x			x			10'	14'	15'
	25	x		x				10'	14'	15'
	25	x			x		10'	12'	14'	15'
	20	x			x			10'	14'	15'
	20			x				10'	14'	15'
	20				x			10'	14'	15'
	20	x		x				12'	16'	17'
	20	x			x			12'	16'	17'
	20 dbl.	x		x				16'	21'	22'
	20 dbl.				x			16'	21'	22'
80 to 120 lb.	25	x			x			12'	16'	17'
	20	x	x			x		12'	16'	17'
	20			x		x		12'	16'	17'
	20 dbl.	x			x			16'	21'	22'

*See partition height table for stud spacing.
†Not allowed.

Courtesy National Gypsum Co.

walls, elevator shafts, and stair wells usually require solid and thick walls, which can be made of all plasterboard at reduced cost of construction. There is no limit to the thickness that can be obtained, depending on the need.

JOINT FINISHING

Flat and inside corner joints are sealed with perforated tape embedded in joint compound and with finishing coats of the

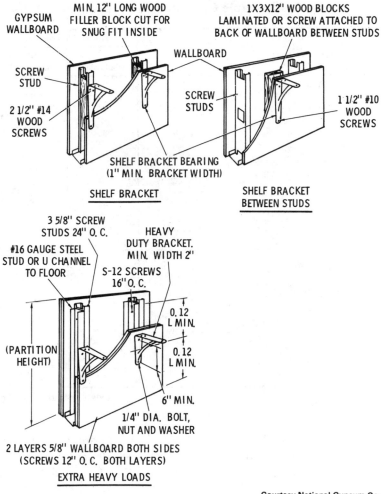

GYPSUM WALLBOARD

MIN. 12" LONG WOOD FILLER BLOCK CUT FOR SNUG FIT INSIDE

1X3X12" WOOD BLOCKS LAMINATED OR SCREW ATTACHED TO BACK OF WALLBOARD BETWEEN STUDS

WALLBOARD

SCREW STUD

2 1/2" #14 WOOD SCREWS

SCREW STUDS

1 1/2" #10 WOOD SCREWS

SHELF BRACKET BEARING (1" MIN. BRACKET WIDTH)

SHELF BRACKET

SHELF BRACKET BETWEEN STUDS

3 5/8" SCREW STUDS 24" O. C.

#16 GAUGE STEEL STUD OR U CHANNEL TO FLOOR

HEAVY DUTY BRACKET. MIN. WIDTH 2"

S-12 SCREWS 16" O. C.

0.12 L MIN.

0.12 L MIN.

(PARTITION HEIGHT)

6" MIN.

1/4" DIA. BOLT, NUT AND WASHER

2 LAYERS 5/8" WALLBOARD BOTH SIDES (SCREWS 12" O. C. BOTH LAYERS)

EXTRA HEAVY LOADS

Fig. 14-28. Recommended method for fastening shelf brackets to plaster-board.

same compound. Outside corners are protected with a metal bead, nailed into place and finished with jointing compound, or a special metal-backed tape, which is also used on inside corners.

Two types of jointing compound are generally available—regular, which takes about 24 hours to dry; and quick-setting,

319

Table 14-2. Allowable Carrying Loads for Anchor Bolts

Type Fastener	Size	Allowable Load, lb	
		½" Wallboard	⅝" Wallboard
HOLLOW WALL	⅛" dia. SHORT	50	—
SCREW ANCHORS	³⁄₁₆" dia. SHORT	65	—
	¼", ⁵⁄₁₆", ³⁄₈" dia. SHORT	65	—
	³⁄₁₆" dia. LONG	—	90
	¼", ⁵⁄₁₆", ³⁄₈" dia. LONG	—	95
COMMON	⅛" dia.	30	90
TOGGLE BOLTS	³⁄₁₆" dia.	60	120
	¼", ⁵⁄₁₆", ³⁄₈" dia.	80	120

Courtesy National Gypsum Co.

which takes about 2½ hours to dry. The plastering mason must make the decision on which type to use, depending on the amount of jointing work he has to do and when he can get around to subsequent coats.

Begin by spotting nail heads with jointing compound. Use the broad knife to smooth out the compound. Apply compound over the joint (Fig. 14-33). Follow this immediately by embedding the perforated tape over the joint (Fig. 14-34). Fig. 14-35 is a closeup view showing the compound squeezed through the perforation of the tape for good keying. Before the compound dries, run the broad knife over the tape to smooth down the compound and to level the surface (Fig.14-36).

After the first coat has dried, apply a second coat (Fig. 14-37) thinly and feather it out 3″ to 4″ on each side of the joint. Also apply a second coat to the nail spots. When the second coat has dried, apply a third coat, also thinly. Feather it out to about 6″ or 7″ from the joint (Fig. 14-38). Final nail spotting is also done at this time.

INSIDE CORNERS

Inside corners are treated in the same way as flat joints, with one exception. The tape must be cut to the proper size and creased down the middle. Apply it to the coated joint (Fig. 14-39), and follow with the treatment mentioned above, but to one side at a time. Let the joint dry before applying the second coat of compound, and do the same for the third coat.

Also available is a corner knife, an L-shaped trowel which allows the installer to smooth the joint compound in one pass.

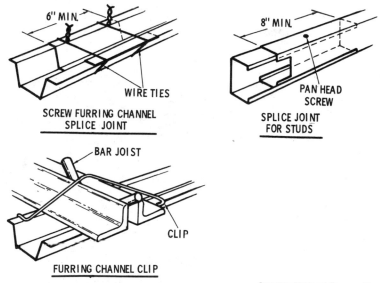

6″ MIN.

WIRE TIES

SCREW FURRING CHANNEL
SPLICE JOINT

8″ MIN.

PAN HEAD
SCREW

SPLICE JOINT
FOR STUDS

BAR JOIST

CLIP

FURRING CHANNEL CLIP

Courtesy National Gypsum Co.

Fig. 14-29. Metal ceiling channels may be fastened to metal joists by wire ties or special clips.

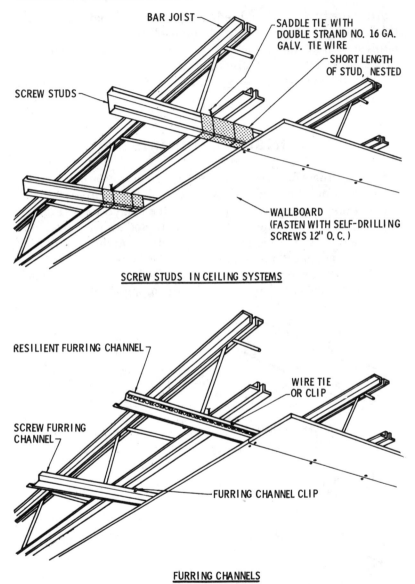

BAR JOIST

SADDLE TIE WITH
DOUBLE STRAND NO. 16 GA.
GALV. TIE WIRE

SHORT LENGTH
OF STUD, NESTED

SCREW STUDS

WALLBOARD
(FASTEN WITH SELF-DRILLING
SCREWS 12" O.C.)

SCREW STUDS IN CEILING SYSTEMS

RESILIENT FURRING CHANNEL

WIRE TIE
OR CLIP

SCREW FURRING
CHANNEL

FURRING CHANNEL CLIP

FURRING CHANNELS

Courtesy National Gypsum Co.

Fig. 14-30. Complete details showing two methods of mounting wallboard
to suspended ceiling structures.

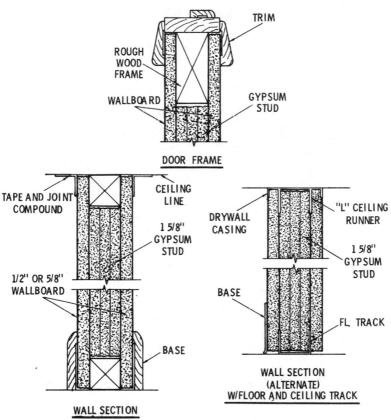

Fig. 14-31. Cross-sectional view of a semisolid, all-plasterboard wall.

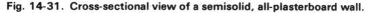

OUTSIDE CORNERS

Outside corners need extra reinforcement because of the harder knocks they may take from the family. Metal corner beads (Fig. 14-40) are used. Nail through the bead into the plasterboard and framing. Apply joint compound over the beading, using a broad knife, as shown in Fig. 14-41. The final treatment is the same as for other joints. The first coat should be about 6″

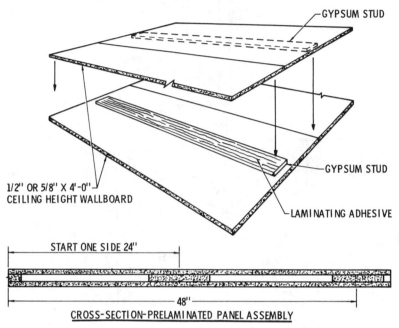

LAMINATING PANEL ASSEMBLIES

GYPSUM STUD

GYPSUM STUD

LAMINATING ADHESIVE

1/2" OR 5/8" X 4'-0"
CEILING HEIGHT WALLBOARD

START ONE SIDE 24"

48"

CROSS-SECTION-PRELAMINATED PANEL ASSEMBLY

Courtesy National Gypsum Co.

Fig. 14-32. Plasterboard laminated together with wide pieces of plaster-board acting as wall studs.

wide, and the second coat about 9″. Feather out the edges, and work the surface smooth with the broad knife and a wet sponge.

PREFINISHED WALLBOARD

Prefinished wallboard is surfaced with a decorative vinyl material. Since it is not painted or papered after installation, in order to maintain a smooth and unmarred finish, its treatment is slightly different than that of standard wallboard. In most installations, nails and screws are avoided (except at top and bottom), and no tape and joint compound are used at the joints.

Fig. 14-33. The first step in finishing joints between plasterboard.

Fig. 14-34. Place perforated tape over the joint and embed it into the joining compound.

325

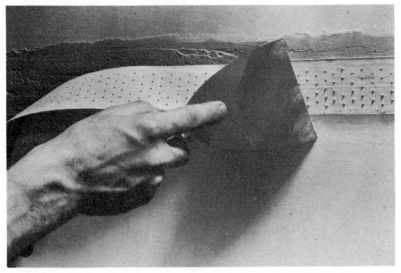

Courtesy National Gypsum Co.

Fig. 14-35. Closeup view showing how a properly embedded tape will show beads of the compound through the perforations.

Courtesy National Gypsum Co.

Fig. 14-36. Using a broad knife to smooth out the compound and to feather edges.

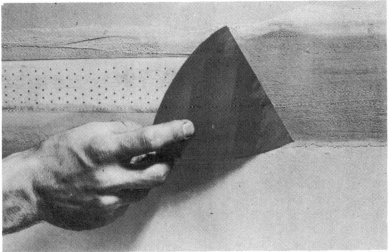

Courtesy National Gypsum Co.

Fig. 14-37. Apply the second coat of compound over perforated tape after the first coat has dried. This coat must be applied smooth with a good feathered edge.

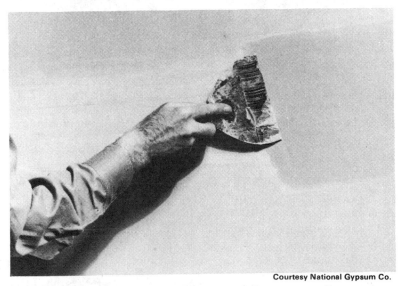

Courtesy National Gypsum Co.

Fig. 14-38. The final coat is applied with a wide knife, carefully feathering the edges. A wet sponge will eliminate any need for further smoothing when dry.

Fig. 14-39. Inside corners are handled in the same manner as the flat joints.

Three basic methods are used to install prefinished wallboard:

1. Nailing to 16″ on center studs or to furring strips, but using colored and matching nails available for the purpose.
2. Cementing to the studs or furring strips with adhesive, nailing only at the top and bottom. These nails may be matching colored or plain nails, which are covered with matching cove molding and base trim.
3. Laminating to a base layer on regular wallboard, or to old wallboard in the case of existing wall installations.

Before installing prefinished wallboard, a careful study of the wall arrangement should be made. Joints should be centered on architectural features such as fireplaces, windows, etc. End panels should be of equal width. Avoid narrow strips as much as possible.

Decorator wallboard is available in lengths to match most wall heights without further cutting. It should be installed vertically and should be about ⅛" shorter than the actual height so that it will not be necessary to force it into place. Prefinished wallboards have square edges and are butted together at the joints, with or without the vinyl surface lapped over the edges.

As with standard wallboard, prefinished wallboard is easily cut into a narrow piece by scoring and snapping. Place the board on a flat surface with the vinyl side up. Score the vinyl side with a dimension about 1" wider than the width of the panel (Fig. 14-42). Turn the vinyl surface face down and score the back edge to

Courtesy National Gypsum Co.

Fig. 14-40. Outside corners are generally reinforced with a metal head. It is nailed in place going through the plasterboard into the stud.

329

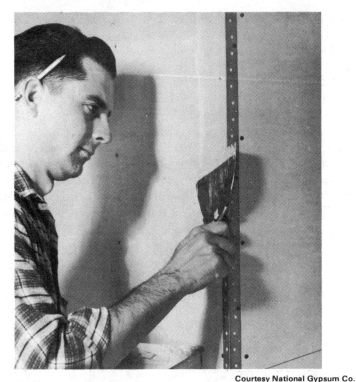

Courtesy National Gypsum Co.

Fig. 14-41. Applying compound over the corner head.

the actual dimension desired. In both cases use a good straight board as a straightedge. Place the board over the edge of a long table, and snap the piece off (Fig. 14-43). This will leave a piece of vinyl material hanging along the edge. Fold the material back, and tack it into place onto the back of the wallboard (Fig. 14-44).

Cutouts are easily made on wallboard with a fine-toothed saw. Where a piece is to be cut out, as for a window, saw along the narrower cuts, and then score the longer dimension and snap off the piece (Fig. 14-45). Circles are cut out by first drilling a hole large enough to insert the end of the keyhole saw (Fig. 14-46). Square cutouts for electrical outlet boxes need not be sawed but can be punched out. Score through the vinyl surface, as shown in Fig. 14-47. Give the area a sharp blow and it will break through.

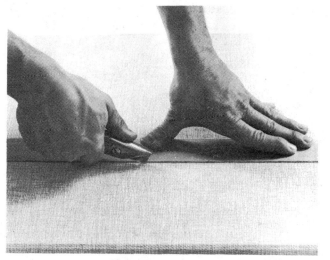

Courtesy National Gypsum Co.

Fig. 14-42. When cutting vinyl-finished wallboard, the cut should be about 1" wider than the desired width. This is to allow an inch of extra vinyl for edge treatment.

Courtesy National Gypsum Co.

Fig. 14-43. Score the back of the wallboard to the actual width desired, leaving a 1" width of vinyl.

331

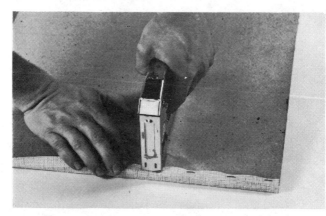

Fig. 14-44. Fold the 1" vinyl material over the edge of the wallboard and tack in place. This will provide a finished edge without further joint treatment.

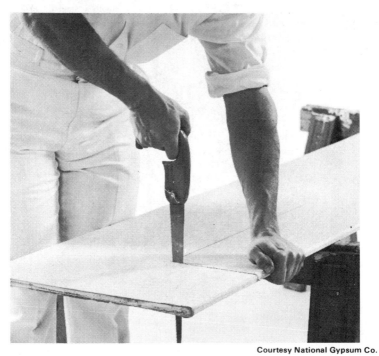

Courtesy National Gypsum Co.

Fig. 14-45. Cutting vinyl-covered wallboard for various openings.

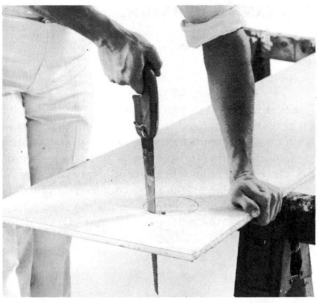

Courtesy National Gypsum Co.

Fig. 14-46. Circle cutouts are made by first drilling a hole large enough to insert a keyhole saw.

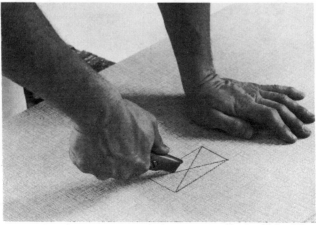

Courtesy National Gypsum Co.

Fig. 14-47. For rectangular cutouts, score deeply on the vinyl side for the outline of the cutout. Tap the section to be removed.

333

INSTALLING PREFINISHED WALLBOARD

Prefinished wallboard may be nailed to studs or furring strips with decorator-type or colored nails. In doing so, however, the job must be done carefully so that the nails make a decorative pattern. Space them every 12″ and not less than ⅜″ from the panel edges (Fig. 14-48). When nailing directly to the studs, be sure the studs are straight and flush. If warped, they may require shaving down at high spots or shimming up at low spots. To avoid extra work, carpenters on new construction should select the best 2″ × 4″ lumber and do a careful job of placing it. Where studs are already in place, it may be easier to install furring strips horizontally and shim them during installation (Fig. 14-49). Use 1″ × 3″ wood, and space them 16″ apart. Over an existing

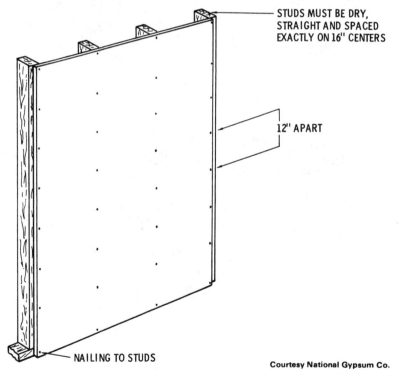

STUDS MUST BE DRY, STRAIGHT AND SPACED EXACTLY ON 16″ CENTERS

12″ APART

NAILING TO STUDS

Courtesy National Gypsum Co.

Fig. 14-48. Prefinished walls may be nailed directly to studs as shown.

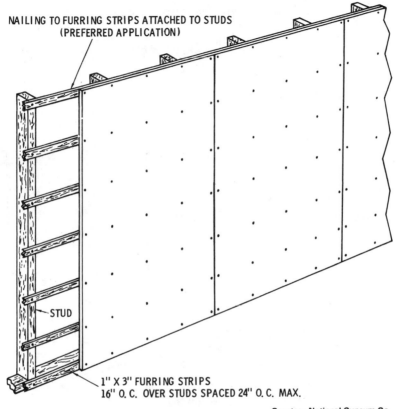

NAILING TO FURRING STRIPS ATTACHED TO STUDS
(PREFERRED APPLICATION)

STUD

1" X 3" FURRING STRIPS
16" O. C. OVER STUDS SPACED 24" O. C. MAX.

Fig. 14-49. Wood furring strips may be installed over wall studs that are not straight.

broken plaster wall or solid masonry, install furring strips as shown in Fig. 14-50. Use concrete nails to fasten the furring strips on solid masonry.

The most acceptable method of installing prefinished wallboard is with adhesive, either to vertical studs or horizontal furring strips, or to existing plasterboard or solid walls. In order to get effective pressure over the entire area of the boards for good adhesion, the panels must be slightly bowed. This is done in the way shown in Fig. 14-51. A stock of wallboard is placed either face down over a center support or face up across two end supports.

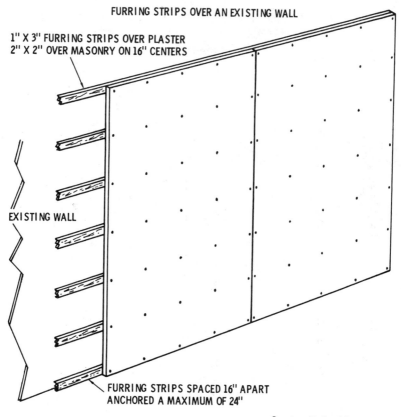

FURRING STRIPS OVER AN EXISTING WALL

1" X 3" FURRING STRIPS OVER PLASTER
2" X 2" OVER MASONRY ON 16" CENTERS

EXISTING WALL

FURRING STRIPS SPACED 16" APART
ANCHORED A MAXIMUM OF 24"

Fig. 14-50. Wood furring strips installed over solid masonry or old existing walls.

The supports can be a 2″ × 4″ lumber, but if they are in contact with the vinyl finish, they must be padded to prevent marring. It may take one day or several days to get a moderate bow, depending on the weather and humidity.

The adhesive may be applied directly to the studs (Fig. 14-52) or to furring strips (Fig. 14-53). Where panel edges join, run two adhesive lines on the stud, one for each edge of a panel. These lines should be as close to the edge of the 2″ × 4″ stud as possible to prevent the adhesive from oozing out between the panel

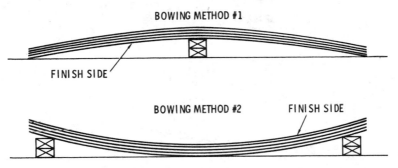

BOWING METHOD #1

FINISH SIDE

BOWING METHOD #2   FINISH SIDE

Courtesy National Gypsum Co.

Fig. 14-51. Bowing wallboard for the adhesive method of installation.

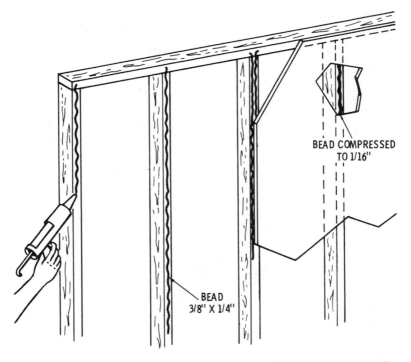

BEAD COMPRESSED
TO 1/16"

BEAD
3/8" X 1/4"

Courtesy National Gypsum Co.

Fig. 14-52. Adhesive applied directly to the wall studs. The adhesive is applied in a wavy line the full length of the stud.

LEAVE 1" SPACE AT
JOINT OF PANELS

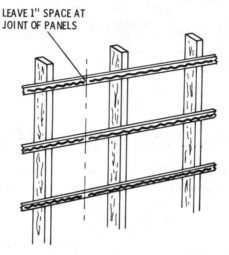

Courtesy National Gypsum Co.

Fig. 14-53. Adhesive applied directly to the wood furring strips.

joints. For the same reason, leave a space of 1" with no adhesive along the furring strips, where panel edges join.

Place the panels in position with a slight sliding motion, and nail the top and bottom edges. Be sure long edges of each panel are butted together evenly. Nail to sill and plate at top and bottom only. The bowed panels will apply the right pressure for the rest of the surface.

The top and bottom nails may be matching colored nails or finished with a cove and wall trim. Special push-on trims are available which match the prefinished wall. Inside corners may be left with panels butted, but one panel must overlap the other. Outside corners must include solid protection. For both inside corners, if desired, and for outside corners, snap-on trim and bead matching the panels can be obtained (Fig. 14-54). They are applied by installing retainer strips first (vinyl for inside corners and steel for outside corners), which hold the finished trim material in place.

Prefinished panels may be placed over old plaster walls or solid walls by using adhesive. If placed on old plaster, the surface must be clean and free of dust or loose paint. If wallpapered, the

Courtesy National Gypsum Co.

Fig. 14-54. Snap-on matching trim is available for prefinished wallboard.

Courtesy National Gypsum Co.

Fig. 14-55. Applying adhesive to prefinished wallboard to be installed over old plaster walls.

339

paper must be removed and walls completely washed. The new panels are bowed, as described before, and then lines of adhesive are run down the length of the panels about 16″ apart (Fig. 14-55). Keep edge lines ¾″ to 1″ from the edges. Apply the boards to the old surface with a slight sliding motion, and fasten at top and bottom with 6d nails or matching colored nails. If boards will not stay in proper alignment, add more bracing against the surface and leave for 24 hours. The top and bottom may be finished with matching trim, as explained before.

CHAPTER 15

Stone and Rock Masonry

With most of the crust of the earth composed of stone or rock (or rocklike materials), it is no wonder there is a large amount of this material used in masonry construction. It is probably one of the oldest construction materials known to man. The earliest solid dwellings were natural caves. The cliff dwellings of the Southwest Indians date back thousands of years. Their homes were hand-hewn out of soft rock material on the sides of the cliffs. The abundance of rock made it a natural choice for early man to pick up loose pieces and pile them rock upon rock to build a wall against enemy invasion or to keep out wild animals. Such uses led to cutting rock into more usable shapes, which then could be stacked more precisely and more substantially into buildings, tombs, pyramids, etc.

The current use of stones and rocks for construction is usually based on:

1. The abundance of the material in the area.
2. The desire to attain a rustic appearance.

More commonly found abroad than in this country are fences of rock around a ranch or farm made from the rocks cleared from the land. They are simply piled on the ground without mortar. Because of the tremendous amount of hand labor involved, modern farms and ranches in this country find it more economical to construct post and wire fences. Rock and stone are carefully chosen, cut and polished to a luster, and used for home and building front facings.

TYPES OF STONE

There are roughly two types of stone. One type is granite and basalts, which is called siliceous. They are hard, usually round in shape, and found in great abundance on the earth's surface. They are usually cut or split into usable shapes by machine, but many are found with flat surfaces on one or two sides and, by careful selection, may be used as is for wall construction. The term used when building with this type of stone is *rubble construction* (Fig. 15-1).

Limestone, shale, and sandstone are the other type of stone. They are stratified in their natural form and are therefore easy to split into flat pieces. They are considered calcareous argillacious (claylike) materials. While this type of stone is easier to work with, it does have one drawback. Because of its stratified makeup, it is subject to earlier deterioration in climates of extreme cold and wetness. Moisture can get into the crevices and, on freezing, break away edge pieces a little at a time. Building with this type of stone is called *ashlar construction*.

KINDS OF STONE

Table 15-1 lists the weight and strength of various types of stone. It should be noted that the figures for strength are only approximate. Stones of the same kind from different parts of the country can vary considerably in strength. For those interested in a more detailed description of the makeup of various stone types, the following will be useful:

Limestone—Limestones consist chiefly of calcium carbonate with small proportions of other substances. They are often classified under four heads: compact limestones consist of carbonate of lime, either pure or in combination with clay and sand. Granular, or oolitic, limestones consist of grains of carbonate clay. The

(A) Random bond.

(B) Coarse bond.

Fig. 15-1. Two kinds of rubble construction.

Table 15-1. Approximate Weight and Strength of Stone

Kind of Stone	Weight, lb. per cu. ft.	Crushing Strength, psi	Shearing Strength, psi
Sandstone	150	8,000	1,500
Granite	170	15,000	2,000
Limestone	170	6,000	1,000
Marble	170	10,000	1,400
Slate	175	15,000	
Trap Rock	185	20,000	

grains are egg-shaped (hence the name "oolite") and vary in size from tiny particles to grains as large as peas. Shelly limestones consist almost entirely of small shells, cemented together by carbonate of lime. Magnesian limestones are composed of carbonates of lime and magnesia in varying proportions, and usually also contain small quantities of silica, iron, and aluminum. The hardest and closest grained of these are capable of taking a fine polish. Limestones should be used with care since they are uncertain in behavior and usually more difficult to work than sandstones. As a general rule, they do not withstand fire well.

Sandstone—Sandstones are composed of grains of sand held together by a cementing substance to form a compact rock. The cementing medium may be silica, aluminum, carbonate of lime, or an oxide of iron. Those stones that have a siliceous cement are the most durable. Sandstones vary more in color than limestones, the color being largely due to the presence of iron. Cream, brown, gray pink, red, light and dark blue, and drab are common colors. The texture of sandstones varies from a fine, almost microscopic, grain to one composed of large particles of sand. It will generally be found that the heaviest, densest, least porous, and most lasting stones are those with a fine grain.

Granite—Granites are igneous rock formed by volcanic action and are of all geological ages. Granite is composed of quartz, felspar, and mica intimately compacted in varying proportions to form a hard granular stone. Quartz is the principal constituent and imparts to the rock the qualities of durability and strength. Stones containing a large proportion of quartz are hard and difficult to work. Felspar of an earthly nature is opaque and is liable to decay; it should be clear and almost transparent. The characteristic color of the granite is generally due to this substance, but the stone is often affected by the nature of the mica it contains, whether it is light or dark in tint. Granite is the hardest, strongest, and most durable of building stones and is difficult and costly to work. When polished, many varieties present a beautiful and lasting surface. By reason of its strength and toughness this stone is often used for foundations, bases, columns, curbs, and paving and in all positions where great strength is required.

Slate—Slate used for roofing and other purposes in building is a fine-grained and compact rock composed of sandy clay which has been more or less metamorphosed by the action of heat and tremendous pressure. Such rocks were originally deposited in the form of sediment by a sea or river, afterward becoming compacted by the continual heaping up of superincumbent material. Owing, no doubt, to a sliding motion that took place at some time, slatey rocks are capable of being split into thin sheets, which are trimmed to the various marketable sizes. A good slate is hard, tough, and nonabsorbent. It will give out a metallic ring if struck, and when trimmed it will not splinter, nor will the edges become ragged. Slates range in color from purple to gray and green.

ROCK AND STONE USES

Rocks and stones are used for construction or for facing to achieve a rustic appearance (Fig. 15-2). Construction is usually confined to boundary walls or fencing (Fig. 15-3), earth retaining walls (Fig. 15-4), dams, etc. Complete housing or building structures are seldom made of stone exclusively. Veneer stone is mortar-set against an existing building and adds no support to the structure (Fig. 15-5). Various color and texture combinations can be made by careful selection of polished stone (Fig. 15-6). Facing stone may be small round or flat pieces, usually highly polished. The small round stones are carefully selected for uniformity of size and shape. Flat rock is generally factory-cut and polished. It is flat, with little depth, and must be set in with great care as to pattern to retain a flat surface and for cleanliness of the finished work.

Fig. 15-2. The stone facing of this wall has the appearance of solid-rock construction, but the facing stones are shallow and used for a rustic western look.

Fig. 15-3. An expertly built stone wall used for fencing between home lots. Note the random rubble type of construction capped with a layer of brick.

Fig. 15-4. Because of the slope of the land on which this home is built, some form of retaining wall is needed to prevent soil erosion during rain. Natural stone, carefully selected and cut, forms a retaining wall which is both attractive and functional.

Fig. 15-5. The entrance to a restaurant with walls veneered with flat-cut and polished stone.

Fig. 15-6. By careful selection and polishing, the stones are made to vary in color and texture.

Most solid structures are in the form of walls, either self-supporting or for earth-retaining. Either rubble or ashlar construction is used, with ashlar predominating because of the ease with which the rock is split. Rubble construction requires careful selection of rocks with flat sides or purchasing the rock with factory-cut sides. Concrete block can also be purchased with one side treated to look like natural rock. Such blocks are easier to lay up and require much less mortar than natural stone.

Rocks may be laid up dry or with mortar, with the latter preferred. While dry construction is somewhat obsolete, there is one example of modern dam construction using no mortar. In Ciudad de Juarez, Mexico, a water-retaining dam was built with stone so carefully cut and laid that no mortar was used, and there is no leak in the dam. Mortar construction is generally the same as for laying up brick or concrete blocks. The mortar ingredients are about the same. A considerable amount of mortar is usually required because of the irregularity of the stone faces. The experienced mason or homeowner, if he builds his own, will fill voids between rocks with smaller stones as much as possible to reduce the amount of mortar needed; the stones are stronger than the mortar.

CONSTRUCTION CONSIDERATIONS

With the exception of stones laid on the ground without mortar, stone walls or rough fencing require a firm foundation, preferably of concrete, as with brick or concrete wall construction. Without a solid foundation, heaving of the earth as a result of freezing and thawing will crack the wall and require frequent repairs. (See Volume 1 of this library for detailed instructions on footing construction.) In general, a trench should be dug below the frost line and have a width about twice that of the wall to be built. Pour concrete up to just below the level of the ground. Allow it to cure for 48 hours before beginning the stone wall construction.

Determine the pattern desired and the construction method to be used, and then order stones or rocks to fit. Random rubble masonry has the most rustic appearance but may require more care in setting up a desirable pattern. Ashlar construction begins to approach the system of bricklaying in that there is some resemblance to course-upon-course construction. Halfway between these two is the random-coursed rubble construction, probably the most popular pattern used. The stones generally have a somewhat flat face, but the nearly straight mortar lines identifying the courses are altered (Fig. 15-7).

A proportion of the stones should be long enough to reach from the front of the wall to the back. These stones will act as headers, as with brick, and are necessary for good wall strength (Fig. 15-8). Stones should be small enough for easy handling. Considering that a piece only 1 cu. ft. in size weighs over 150 lb., hand-laying makes size selection important. Where larger sizes are specified, contractors should place the stones with power equipment, using rock-handling tongs. For random masonry, a quantity of small stones should be used to fill between large spaces and save on the cost of mortar.

In constructing an earth retaining wall, there must be a backfill of stone and a drainpipe properly placed (Fig. 15-9). It is impor-

Fig. 15-7. Random-course rubble wall construction. It takes carefully cut stones to make mortar lines as fine as shown here.

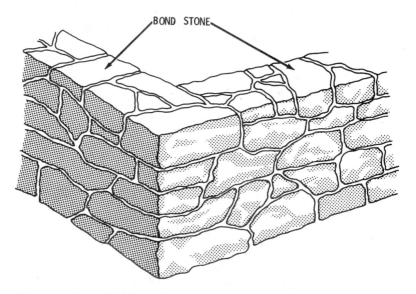

Fig. 15-8. The use of long stones as headers to bond the wall transversely.

Fig. 15-9. Loose rock and a drainpipe were installed behind this stone retaining wall before the earth was filled in.

tant to provide water drainage to prevent moisture from collecting behind the wall, which could seep into the mortar. This could weaken the mortar and cause crumbling. All stones and rocks should be washed and allowed to drain dry before using. Any dust on the stones prevents proper cementing of the mortar to the stone. All necessary material should be on hand before starting to lay a stone wall—tools, stones of all sizes, and plenty of mortar ingredients. Remember, much more mortar will be used for stone walls than for brick or concrete-block masonry.

LAYING A STONE-MASONRY WALL

The strongest mortar is regular sand and concrete, consisting of 1 part portland cement and 3 parts sand, plus water. However, this mortar does have some shrinkage and locks the workability of mortar cement. For ease of laying, use a portland-cement–lime–sand mix, the same as that used for laying bricks. Regular portland cement tends to stain stones and rocks; this effect can be lessened by the use of white portland cement in the mortar.

351

Fig. 15-10. Nearly all stones have a cleavage line. Stones are more easily broken, when necessary, by breaking with a chisel at the cleavage line.

With the foundation installed and all materials and tools on hand, lay a bed of mortar over the foundation to take the first course of stones. The bed of mortar must be thick to fill all crevices of the bottom stones. If necessary, include small stones. Each stone should be laid with its broadest face horizontal. Place the larger stones for the bottom courses, both for strength and better appearance. Porous stones should be dampened before placing to prevent too much water absorption form the mortar. If it is necessary to reset a stone, lift it clear of its position and reset with new mortar. Remember the importance of good bonding. Place long stretcher stones about every 6 to 10 sq. ft. of wall. Offset adjacent header stones above and below them.

If alignment of the wall is not important, the courses of stone may be laid and aligned by sight. For precise alignment, drive a stake at each end of the wall and stretch a string across as a gauge for a straight line. Move the string up the stakes as the courses are laid. The top course of stones should present a flat appearance. This may require some readjustment of the course below to fit the top course. Save flat-faced stones for the top course.

It is frequently necessary to cut some stones on the job. Nearly all stones have cleavage lines. By placing a large chisel at the cleavage line, a few strokes of a small sledge should break the stone at or near this line, as shown in Fig. 15-10. A mason's hammer may be used to chip away protruding pieces from stones (Fig. 15-11). Wear face goggles when doing this.

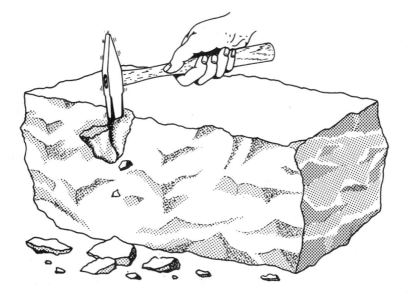

Fig. 15-11. A mason's hammer used on the job to break away protruding lines.

Finally, tool the mortar joints with a stick or jointing tool. Clean away excess mortar. Use a wet sponge to clean off the face of stones which have mortar spots. If mortar is left, it will be difficult to remove when hard and may stain the stone.

CHAPTER 16

How to Read Blueprints

Whether the mason is his own contractor, is working for a contractor, or is an apprentice, it is important that he know how to read and follow a blueprint for the structure he is working on. In order to do his job properly, it is not necessary for the mason to know how to draw a blueprint, but only to read one. The mason must know what work he is to do on a structure and where it is to be done.

Blueprints are made by the architect of the structure or by members of his staff. In some cases the architect oversees the work itself, or it is turned over to a general contractor. In any case, there is someone with whom the mason can communicate if there are any questions about details. Blueprint details and symbols have been standardized, and, once understood, there is usually no problem in proceeding with the work required.

DEVELOPING THE BLUEPRINT

Start with a small block of wood, 1″ × 1″ × 4″. Look at it, and compare it with the top illustration in Fig. 16-1. The illustration is called an *orthographic view* of the object. It differs from an actual view in that parallel lines are drawn truly parallel, whereas when the object itself is viewed, the parallel lines (AB, EF, DC, and HG) seem to taper to a vanishing point. This is the same effect as when the object is photographed. While an orthographic

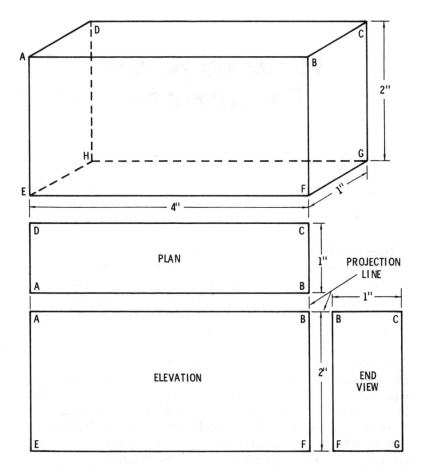

Fig. 16-1. Projecting an orthographic view of a wood block to flat planes.

view is not as actually seen, it does permit the use of equal dimensions where such dimensions exist.

The lower part of Fig. 16-1 is a projection of the orthographic view, in which each plane of the object is shown as flat illustrations on a sheet of paper. The elevation view is the height and width of the object, and the end view is as named. The *plan view* is looking straight down on the object.

Fig. 16-2 is an equivalent drawing, but of a simple structure such as a barn. The plan view shown here is that of the roof of the structure. The plan view could also be the floor line of he struc-

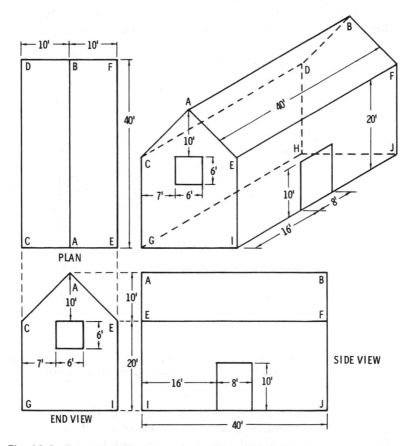

Fig. 16-2. Steps in projecting an orthographic view of a barn to plan views. In construction blueprints, the plan view is the most important.

357

ture by omitting the line AB. It is the *plan view* that is used most in architectural blueprints, and it is in this view that other details are included. In a multistory structure, there would be a plan view of each floor. The drawing would include details of equipment to be supplied by the contractor or by the subcontractors, such as mason contractors, electrical contractors, plumbing contractors, etc.

The architect will discuss a structure with the customer, and will make some rough hand drawings during the discussion. Back at his office, the hand sketches will be made to look a little more formal but will not include dimensions, equipment, or electrical and plumbing details. They will locate the rooms, doors, windows, stairs, and other important features (Fig. 16-3).

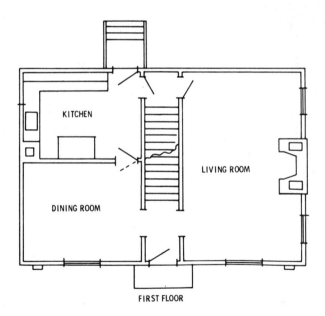

Fig. 16-3. The first ruled drawings made by an architect show essential features, leaving out many details until they are finalized with the customer.

Further discussions will finalize dimensions showing equipment and other considerations. The architect will then return to the drawing board to make the final blueprints. Fig. 16-4 shows a

358

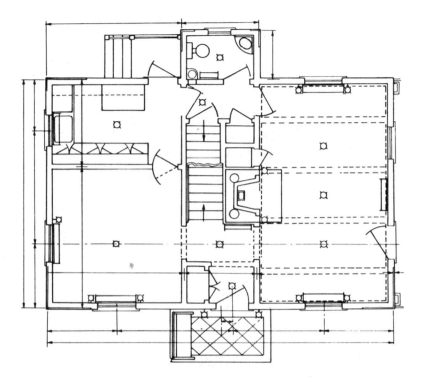

Fig. 16-4. Plan view of the first floor of a residence. All details are now shown, except dimensions.

print with everything included except dimensions, which are promptly added. It is from a print like this, with dimensions, that contractors do their work. Fig. 16-5 shows a floorplan of a basement of a home, with all dimensions shown.

A set of blueprints for a home or building consists of many drawings. There will be elevations views, end views, and plan views for each floor, as well as the roof. In addition, special structural details will be shown separately but keyed to the main plan view. Fig. 16-6 shows the details of a basement wall, footing, and floor. This print is used by the excavator and concrete mason. Fig. 16-7 shows details of the wall construction. It is used by the brick mason, the plasterer, and the carpenter.

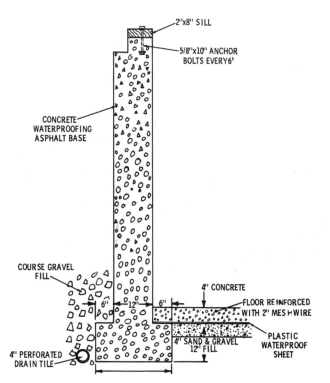

Fig. 16-5. Plan view of a basement, now including dimensions.

SCALED DIMENSIONS

All blueprints are drawn to scale; that is, ½″ of print is equal to 1 ft. of actual construction, or some other ratio. The scale ratio is given on the print. Should a dimension be questioned, it is easy to check with an architect's scale ruler. The one shown in Fig. 16-8 has three edges; it has 11 proportionate scales, from 3″ to 1 ft. down to ³/₃₂″ to 1 ft., in addition to a standard 12″ rule. Many scales are obtained by reading from both ends of the ruler. The beginning end of each scale includes scaled inches.

360

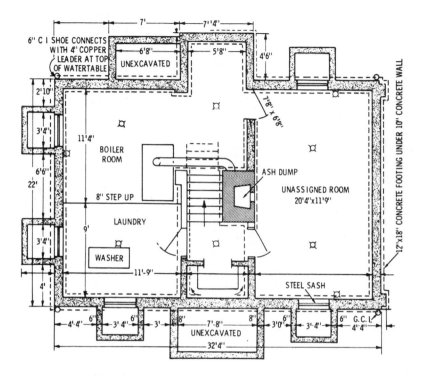

Fig. 16-6. Details of a basement wall, floor, and footing.

STANDARD SYMBOLS

The symbols used by architects to represent details in a blueprint have become standardized, and masons must be acquainted with them. The use of symbols prevents the cluttering of prints with small details that may not show clearly.

Fig. 16-9 shows the standard symbols pertaining to walls and openings that are important to masons. The distance between parallel lines is the scaled thickness of the walls. The actual construction details would be shown in separate drawings, as menti-

361

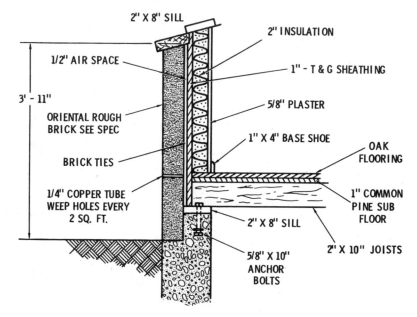

2" X 8" SILL

2" INSULATION

1/2" AIR SPACE

1" - T & G SHEATHING

3' - 11"

5/8" PLASTER

ORIENTAL ROUGH
BRICK SEE SPEC

1" X 4" BASE SHOE

OAK
FLOORING

BRICK TIES

1/4" COPPER TUBE
WEEP HOLES EVERY
2 SQ. FT.

1" COMMON
PINE SUB
FLOOR

2" X 8" SILL

5/8" X 10"
ANCHOR
BOLTS

2" X 10" JOISTS

Fig. 16-7. Details of floor. and inside and outside of wall construction.

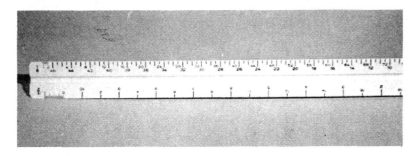

Fig. 16-8. An architect's rule, with fractions of an inch to a foot.

oned above, except where such construction is assumed standard
in the industry and dictated by building codes, either national or
local.

Fig. 16-10 shows the standard symbols dictating the material to
be used in construction. These have been agreed upon by the

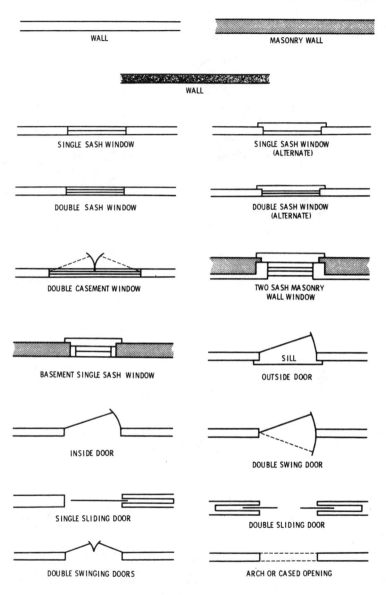

WALL

MASONRY WALL

WALL

SINGLE SASH WINDOW

SINGLE SASH WINDOW
(ALTERNATE)

DOUBLE SASH WINDOW

DOUBLE SASH WINDOW
(ALTERNATE)

DOUBLE CASEMENT WINDOW

TWO SASH MASONRY
WALL WINDOW

BASEMENT SINGLE SASH WINDOW

SILL

OUTSIDE DOOR

INSIDE DOOR

DOUBLE SWING DOOR

SINGLE SLIDING DOOR

DOUBLE SLIDING DOOR

DOUBLE SWINGING DOORS

ARCH OR CASED OPENING

Fig. 16-9. Standard symbols used for walls and openings.

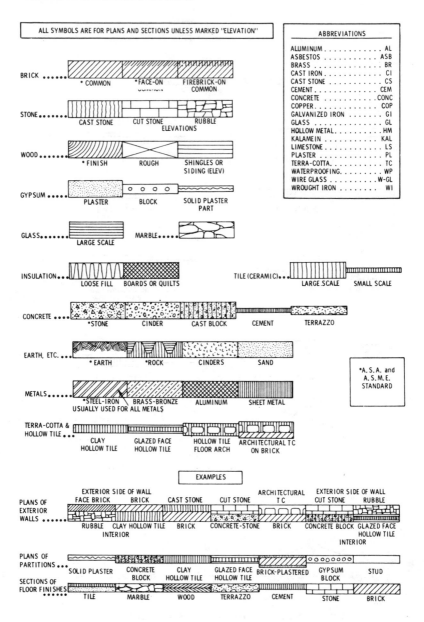

Fig. 16-10. Standard symbols for the material in walls and floors.

ASA and ASME, societies which do testing and have set standards for the building trades.

SPECIFICATIONS

By definition, a *specification* is a definite, particularized, and complete statement setting forth the nature and construction of the object to which it relates. As applied to the building trades, specifications describe briefly, yet exactly, each item in a list of the materials required to complete a contract for building the entire project.

Great care should be used when writing specifications to avoid misunderstandings. Each item entering into the construction should be defined and described with such precision that there can be no chance of misunderstanding or double interpretation. The language should be simple and brief. For the guidance of architects in writing specifications, the American Institute of Architects has prepared a number of standard documents, and these should be carefully studied and consulted by the architect when writing specifications for building and constructing any size project.

Specifications should refer to the contract form of which they are to become a part. This saves repetition of statements with regard to liability of contractor, owner, etc.

Index

INDEX

368

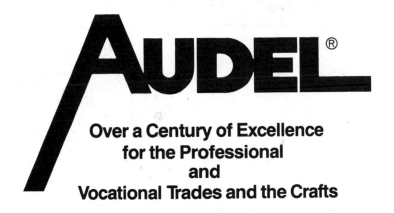

**Over a Century of Excellence
for the Professional
and
Vocational Trades and the Crafts**

**Order now from your local bookstore
or use the convenient order form at
the back of this book.**

AUDEL

These fully illustrated, up-to-date guides and manuals mean a better job done for mechanics, engineers, electricians, plumbers, carpenters, and all skilled workers.

Contents

Fractional Horsepower Electric Motors

Rex Miller and Mark Richard Miller
5½ x 8¼ Hardcover 436 pp. 285 illus.
ISBN: 0-672-23410-6 $15.95

Fully illustrated guide to small-to-moderate-size electric motors in home appliances and industrial equipment: • terminology • repair tools and supplies • small DC and universal motors • split-phase, capacitor-start, shaded pole, and special motors • commutators and brushes • shafts and bearings • switches and relays • armatures • stators • modification and replacement of motors.

Electrical

House Wiring sixth edition
Roland E. Palmquist
5½ x 8 ¼ Hardcover 256 pp. 150 illus.
ISBN: 0-672-23404-1 $13.95

Rules and regulations of the current National Electrical Code® for residential wiring, fully explained and illustrated: • basis for load calculations • calculations for dwellings • services • nonmetallic-sheathed cable • underground feeder and branch-circuit cable • metal-clad cable • circuits required for dwellings • boxes and fittings • receptacle spacing • mobile homes • wiring for electric house heating.

Practical Electricity fourth edition
Robert G. Middleton; revised by L. Donald Meyers
5½ x 8¼ Hardcover 504 pp. 335 illus.
ISBN: 0-672-23375-4 $14.95

Complete, concise handbook on the principles of electricity and their practical application: • magnetism and electricity • conductors and insulators • circuits • electromagnetic induction • alternating current • electric lighting and lighting calculations • basic house wiring • electric heating • generating stations and substations.

Guide to the 1984 Electrical Code®
Roland E. Palmquist
5½ × 8¼ Hardcover 664 pp. 225 illus.
ISBN: 0-672-23398-3 $19.95

Authoritative guide to the National Electrical Code® for all electricians, contractors, inspectors, and homeowners: • terms and regulations for wiring design and protection • wiring methods and materials • equipment for general use • special occupancies • special equipment and conditions • and communication systems. Guide to the 1987 NEC® will be available in mid-1987.

Mathematics for Electricians and Electronics Technicians
Rex Miller
5½ x 8¼ Hardcover 312 pp. 115 illus.
ISBN: 0-8161-1700-4 $14.95

Mathematical concepts, formulas, and problem solving in electricity and electronics: • resistors and resistance • circuits • meters • alternating current and inductance • alternating current and capacitance • impedance and phase angles • resonance in circuits • special-purpose circuits. Includes mathematical problems and solutions.

Electric Motors
Edwin P. Anderson; revised by Rex Miller
5½ x 8¼ Hardcover 656 pp. 405 illus.
ISBN: 0-672-23376-2 $14.95

Complete guide to installation, maintenance, and repair of all types of electric motors: • AC generators • synchronous motors • squirrel-cage motors • wound rotor motors • DC motors • fractional-horsepower motors • magnetic contractors • motor testing and maintenance • motor calculations • meters • wiring diagrams • armature windings • DC armature rewinding procedure • and stator and coil winding.

Home Appliance Servicing fourth edition
Edwin P. Anderson; revised by Rex Miller
5½ x 8¼ Hardcover 640 pp. 345 illus.
ISBN: 0-672-23379-7 $15.95

Step-by-step illustrated instruction on all types of household appliances: • irons • toasters • roasters and broilers • electric coffee makers • space heaters • water heaters • electric ranges and microwave ovens • mixers and blenders • fans and blowers • vacuum cleaners and floor polishers • washers and dryers • dishwashers and garbage disposals • refrigerators • air conditioners and dehumidifiers.

Television Service Manual
fifth edition
Robert G. Middleton; revised by Joseph G. Barrile
5½ x 8¼ Hardcover 512 pp.   395 illus.
ISBN: 0-672-23395-9 $15.95

Practical up-to-date guide to all aspects of television transmission and reception, for both black and white and color receivers: • step-by-step maintenance and repair • broadcasting • transmission • receivers • antennas and transmission lines • interference • RF tuners • the video channel • circuits • power supplies • alignment • test equipment.

Electrical Course for Apprentices and Journeymen
second edition
Roland E. Palmquist
5½ x 8¼ Hardcover 478 pp.   290 illus.
ISBN:0-672-23393-2 $14.95

Practical course on operational theory and applications for training and re-training in school or on the job: • electricity and matter • units and definitions • electrical symbols • magnets and magnetic fields • capacitors • resistance • electromagnetism • instruments and measurements • alternating currents • DC generators • circuits • transformers • motors • grounding and ground testing.

Questions and Answers for Electricians Examinations eighth edition
Roland E. Palmquist
5½ x 8¼ Hardcover 320 pp.   110 illus.
ISBN: 0-672-23399-1 $12.95

Based on the current National Electrical Code®, a review of exams for apprentice, journeyman, and master, with explanations of principles underlying each test subject: • Ohm's Law and other formulas • power and power factors • lighting • branch circuits and feeders • transformer principles and connections • wiring • batteries and rectification • voltage generation • motors • ground and ground testing.

Machine Shop and Mechanical Trades

Machinists Library
fourth edition   3 vols
Rex Miller
5½ x 8¼ Hardcover 1,352 pp.   1,120 illus.
ISBN: 0-672-23380-0 $38.95

Indispensable three-volume reference for machinists, tool and die makers, machine operators, metal workers, and those with home workshops.

Volume I, Basic Machine Shop
5½ x 8¼ Hardcover 392 pp.   375 illus.
ISBN: 0-672-23381-9 $14.95

• Blueprint reading • benchwork • layout and measurement • sheet-metal hand tools and machines • cutting tools • drills • reamers • taps • threading dies • milling machine cutters, arbors, collets, and adapters.

Volume II, Machine Shop
5½ x 8¼ Hardcover 528 pp. 445 illus
ISBN: 0-672-23382-7 $14.95

• Power saws • machine tool operations • drilling machines • boring • lathes • automatic screw machine • milling • metal spinning.

Volume III, Toolmakers Handy Book
5½ x 8¼ Hardcover 432 pp.   300 illus.
ISBN: 0-672-23383-5 $14.95

• Layout work • jigs and fixtures • gears and gear cutting • dies and diemaking • toolmaking operations • heat-treating furnaces • induction heating • furnace brazing • cold-treating process.

Mathematics for Mechanical Technicians and Technologists
John D. Bies
5½ x 8¼ Hardcover 392 pp.   190 illus.
ISBN: 0-02-510620-1 $17.95

Practical sourcebook of concepts, formulas, and problem solving in industrial and mechanical technology: • basic and complex mechanics • strength of materials • fluidics • cams and gears • machine elements • machining operations • management controls • economics in machining • facility and human resources management.

Millwrights and Mechanics Guide
third edition
Carl A. Nelson
5½ x 8¼ Hardcover 1,040 pp.   880 illus.
ISBN: 0-672-23373-8 $22.95

Most comprehensive and authoritative guide available for millwrights and mechanics at all levels of work or supervision: • drawing and sketching • machinery and equipment installation • principles of mechanical power transmission • V-belt drives • flat belts • gears • chain drives • couplings • bearings • structural steel • screw threads •.mechanical fasteners • pipe fittings and valves • carpentry • sheet-metal work • blacksmithing • rigging • electricity • welding • pumps • portable power tools • mensuration and mechanical calculations.

Welders Guide third edition
James E. Brumbaugh
5½ x 8 ¼ Hardcover 960 pp.   615 illus.
ISBN: 0-672-23374-6 $23.95

Practical, concise manual on theory, operation, and maintenance of all welding machines: • gas welding equipment, supplies, and process • arc welding equipment, supplies, and process • TIG and MIG welding • submerged-arc and other shielded-arc welding processes • resistance, thermit, and stud welding • solders and soldering • brazing and braze welding • welding plastics • safety and health measures • symbols and definitions • testing and inspecting welds. Terminology and definitions as standardized by American Welding Society.

Welder/Fitters Guide
John P. Stewart
8½ x 11 Paperback 160 pp.   195 illus.
ISBN: 0-672-23325-8 $7.95

Step-by-step instruction for welder/fitters during training or on the job: • basic assembly tools and aids • improving blueprint reading skills • marking and alignment techniques • using basic tools • simple work practices • guide to fabricating weldments • avoiding mistakes • exercises in blueprint reading • clamping devices • introduction to using hydraulic jacks • safety in weld fabrication plants • common welding shop terms.

Sheet Metal Work
John D. Bies
5½ x 8¼ Hardcover 456 pp.   215 illus.
ISBN: 0-8161-1706-3 $17.95

On-the-job sheet metal guide for manufacturing, construction, and home workshops: • mathematics for sheet metal work • principles of drafting • concepts of sheet metal drawing • sheet metal standards, specifications, and materials • safety practices • layout • shear cutting • holes • bending and folding • forming operations • notching and clipping • metal spinning • mechanical fastening • soldering and brazing • welding • surface preparation and finishes • production processes.

III

Power Plant Engineers Guide

third edition
Frank D. Graham; revised by Charlie Buffington
5½ x 8¼ Hardcover 960 pp. 530 illus.
ISBN: 0-672-23329-0 $16.95

All-inclusive question-and-answer guide to steam and diesel-power engines: • fuels • heat • combustion • types of boilers • shell or fire-tube boiler construction • strength of boiler materials • boiler calculations • boiler fixtures, fittings, and attachments • boiler feed pumps • condensers • cooling ponds and cooling towers • boiler installation, startup, operation, maintenance and repair • oil, gas, and waste-fuel burners • steam turbines • air compressors • plant safety.

Mechanical Trades Pocket Manual

second edition
Carl A. Nelson
4 × 6 Paperback 364 pp. 255 illus.
ISBN: 0-672-23378-9 $10.95

Comprehensive handbook of essentials, pocket-sized to fit in the tool box: • mechanical and isometric drawing • machinery installation and assembly • belts • drives • gears • couplings • screw threads • mechanical fasteners • packing and seals • bearings • portable power tools • welding • rigging • piping • automatic sprinkler systems • carpentry • stair layout • electricity • shop geometry and trigonometry.

Plumbing

Plumbers and Pipe Fitters Library

third edition 3 vols
Charles N. McConnell; revised by Tom Philbin
5½x8¼ Hardcover 952 pp. 560 illus.
ISBN: 0-672-23384-3 $34.95

Comprehensive three-volume set with up-to-date information for master plumbers, journeymen, apprentices, engineers, and those in building trades.

Volume 1, Materials, Tools, Roughing-In
5½ x 8¼ Hardcover 304 pp. 240 illus.
ISBN: 0-672-23385-1 $12.95

• Materials • tools • pipe fitting • pipe joints • blueprints • fixtures • valves and faucets.

Volume 2, Welding, Heating, Air Conditioning
5½ x 8¼ Hardcover 384 pp. 220 illus.
ISBN: 0-672-23386-x $13.95

• Brazing and welding • planning a heating system • steam heating systems • hot water heating systems • boiler fittings • fuel-oil tank installation • gas piping • air conditioning.

Volume 3, Water Supply, Drainage, Calculations
5½ x 8¼ Hardcover 264 pp. 100 illus.
ISBN: 0-672-23387-8 $12.95

• Drainage and venting • sewage disposal • soldering • lead work • mathematics and physics for plumbers and pipe fitters.

Home Plumbing Handbook

third edition
Charles N. Nelson
8½ x 11 Paperback 200 pp. 100 illus.
ISBN: 0-672-23413-0 $10.95

Clear, concise, up-to-date fully illustrated guide to home plumbing installation and repair: • repairing and replacing faucets • repairing toilet tanks • repairing a trip-lever bath drain • dealing with stopped-up drains • working with copper tubing • measuring and cutting pipe • PVC and CPVC pipe and fittings • installing a garbage disposals • replacing dishwashers • repairing and replacing water heaters • installing or resetting toilets • caulking around plumbing fixtures and tile • water conditioning • working with cast-iron soil pipe • septic tanks and disposal fields • private water systems.

The Plumbers Handbook

seventh edition
Joseph P. Almond, Sr.
4 × 6 Paperback 352 pp. 170 illus.
ISBN: 0-672-23419-x $10.95

Comprehensive, handy guide for plumbers, pipe fitters, and apprentices that fits in the tool box or pocket: • plumbing tools • how to read blueprints • heating systems • water supply • fixtures, valves, and fittings • working drawings • roughing and repair • outside sewage lift station • pipes and pipelines • vents, drain lines, and septic systems • lead work • silver brazing and soft soldering • plumbing systems • abbreviations, definitions, symbols, and formulas.

Questions and Answers for Plumbers Examinations

second edition
Jules Oravetz
5½ x 8¼ Paperback 256 pp. 145 illus.
ISBN: 0-8161-1703-9 $9.95

Practical, fully illustrated study guide to licensing exams for apprentice, journeyman, or master plumber: • definitions, specifications, and regulations set by National Bureau of Standards and by various state codes • basic plumbing installation • drawings and typical plumbing system layout • mathematics • materials and fittings • joints and connections • traps, cleanouts, and backwater valves • fixtures • drainage, vents, and vent piping • water supply and distribution • plastic pipe and fittings • steam and hot water heating.

HVAC

Air Conditioning: Home and Commercial

second edition
Edwin P. Anderson; revised by Rex Miller
5½ x 8¼ Hardcover 528 pp. 180 illus.
ISBN: 0-672-23397-5 $15.95

Complete guide to construction, installation, operation, maintenance, and repair of home, commercial, and industrial air conditioning systems, with troubleshooting charts: • heat leakage • ventilation requirements • room air conditioners • refrigerants • compressors • condensing equipment • evaporators • water-cooling systems • central air conditioning • automobile air conditioning • motors and motor control.

Heating, Ventilating and Air Conditioning Library

second edition 3 vols
James E. Brumbaugh
5½ x 8¼ Hardcover 1,840 pp. 1,275 illus.
ISBN: 0-672-23388-6 $42.95

Authoritative three-volume reference for those who install, operate, maintain, and repair HVAC equipment commercially, industrially, or at home. Each volume fully illustrated with photographs, drawings, tables and charts.

Volume I, Heating Fundamentals, Furnaces, Boilers, Boiler Conversions
5½ x 8¼ Hardcover 656 pp. 405 illus.
ISBN: 0-672-23389-4 $16.95

• Insulation principles • heating calculations • fuels • warm-air, hot water, steam, and electrical heating systems • gas-fired, oil-fired, coal-fired, and electric-fired furnaces • boilers and boiler fittings • boiler and furnace conversion.

Volume II, Oil, Gas and Coal Burners, Controls, Ducts, Piping, Valves
5½ x 8¼ Hardcover 592 pp. 455 illus.
ISBN: 0-672-23390-8 $15.95

• Coal firing methods • thermostats and humidistats • gas and oil controls and other automatic controls •

IV

ducts and duct systems • pipes, pipe fittings, and piping details • valves and valve installation • steam and hot-water line controls.

Volume III, Radiant Heating, Water Heaters, Ventilation, Air Conditioning, Heat Pumps, Air Cleaners
5 1/2 x 8 1/4 Hardcover 592 pp. 415 illus.
ISBN: 0-672-23391-6 $14.95

• Radiators, convectors, and unit heaters • fireplaces, stoves, and chimneys • ventilation principles • fan selection and operation • air conditioning equipment • humidifiers and dehumidifiers • air cleaners and filters.

Oil Burners fourth edition
Edwin M. Field
5 1/2 x 8 1/4 Hardcover 360 pp. 170 illus.
ISBN: 0-672-23394-0 $15.95

Up-to-date sourcebook on the construction, installation, operation, testing, servicing, and repair of all types of oil burners, both industrial and domestic: • general electrical hookup and wiring diagrams of automatic control systems • ignition system • high-voltage transportation • operational sequence of limit controls, thermostats, and various relays • combustion chambers • drafts • chimneys • drive couplings • fans or blowers • burner nozzles • fuel pumps.

Refrigeration: Home and Commercial second edition
Edwin P. Anderson; revised by Rex Miller
5 1/2 x 8 1/4 Hardcover 768 pp. 285 illus.
ISBN: 0-672-23396-7 $17.95

Practical, comprehensive reference for technicians, plant engineers, and homeowners on the installation, operation, servicing, and repair of everything from single refrigeration units to commercial and industrial systems: • refrigerants • compressors • thermoelectric cooling • service equipment and tools • cabinet maintenance and repairs • compressor lubrication systems • brine systems • supermarket and grocery refrigeration • locker plants • fans and blowers • piping • heat leakage • refrigeration-load calculations.

Pneumatics and Hydraulics

Hydraulics for Off-the-Road Equipment second edition
Harry L. Stewart; revised by Tom Philbin
5 1/2 x 8 1/4 Hardcover 256 pp. 175 illus.
ISBN: 0-8161-1701-2 $13.95

Complete reference manual for those who own and operate heavy equipment and for engineers, designers, installation and maintenance technicians, and shop mechanics: • hydraulic pumps, accumulators, and motors • force components • hydraulic control components • filters and filtration, lines and fittings, and fluids • hydrostatic transmissions • maintenance • troubleshooting.

Pneumatics and Hydraulics fourth edition
Harry L. Stewart; revised by Tom Philbin
5 1/2 x 8 1/4 Hardcover 512 pp. 315 illus.
ISBN: 0-672-23412-2 $15.95

Practical guide to the principles and applications of fluid power for engineers, designers, process planners, tool men, shop foremen, and mechanics: • pressure, work and power • general features of machines • hydraulic and pneumatic symbols • pressure boosters • air compressors and accessories • hydraulic power devices • hydraulic fluids • piping • air filters, pressure regulators, and lubricators • flow and pressure controls • pneumatic motors and tools • rotary hydraulic motors and hydraulic transmissions • pneumatic circuits • hydraulic circuits • servo systems.

Pumps fourth edition
Harry L. Stewart; revised by Tom Philbin
5 1/2 x 8 1/4 Hardcover 508 pp. 360 illus.
ISBN: 0-672-23400-9 $15.95

Comprehensive guide for operators, engineers, maintenance workers, inspectors, superintendents, and mechanics on principles and day-to-day operations of pumps: • centrifugal, rotary, reciprocating, and special service pumps • hydraulic accumulators • power transmission • hydraulic power tools • hydraulic cylinders • control valves • hydraulic fluids • fluid lines and fittings.

Carpentry and Construction

Carpenters and Builders Library
fifth edition 4 vols
John E. Ball; revised by Tom Philbin
5 1/2 x 8 1/4 Hardcover 1,224 pp. 1,010 illus.
ISBN: 0-672-23369-x $39.95
Also available in a new boxed set at no extra cost:
ISBN: 0-02-506450-9 $39.95

These profusely illustrated volumes, available in a handsome boxed edition, have set the professional standard for carpenters, joiners, and woodworkers.

Volume 1, Tools, Steel Square, Joinery
5 1/2 x 8 1/4 Hardcover 384 pp. 345 illus.
ISBN: 0-672-23365-7 $10.95

• Woods • nails • screws • bolts • the workbench • tools • using the steel square • joints and joinery • cabinetmaking joints • wood patternmaking • and kitchen cabinet construction.

Volume 2, Builders Math, Plans, Specifications
5 1/2 x 8 1/4 Hardcover 304 pp. 205 illus.
ISBN: 0-672-23366-5 $10.95

• Surveying • strength of timbers • practical drawing • architectural drawing • barn construction • small house construction • and home workshop layout.

Volume 3, Layouts, Foundations, Framing
5 1/2 x 8 1/4 Hardcover 272 pp. 215 illus.
ISBN: 0-672-23367-3 $10.95

• Foundations • concrete forms • concrete block construction • framing, girders and sills • skylights • porches and patios • chimneys, fireplaces, and stoves • insulation • solar energy and paneling.

Volume 4, Millwork, Power Tools, Painting
5 1/2 x 8 1/4 Hardcover 344 pp. 245 illus.
ISBN: 0-672-23368-1 $10.95

• Roofing, miter work • doors • windows, sheathing and siding • stairs • flooring • table saws, band saws, and jigsaws • wood lathes • sanders and combination tools • portable power tools • painting.

Complete Building Construction
second edition
John Phelps; revised by Tom Philbin
5 1/2 x 8 1/4 Hardcover 744 pp. 645 illus.
ISBN: 0-672-23377-0 $19.95

Comprehensive guide to constructing a frame or brick building from the footings to the ridge: • laying out building and excavation lines • making concrete forms and pouring fittings and foundation • making concrete slabs, walks, and driveways • laying concrete block, brick, and tile • building chimneys and fireplaces • framing, siding, and roofing • insulating • finishing the inside • building stairs • installing windows • hanging doors.

Complete Roofing Handbook
James E. Brumbaugh
5½ x 8¼   Hardcover   536 pp.   510 illus.
ISBN: 0-02-517850-4 $29.95

Authoritative text and highly detailed drawings and photographs,on all aspects of roofing: • types of roofs • roofing and reroofing • roof and attic insulation and ventilation • skylights and roof openings • dormer construction • roof flashing details • shingles • roll roofing • built-up roofing • roofing with wood shingles and shakes • slate and tile roofing • installing gutters and downspouts • listings of professional and trade associations and roofing manufacturers.

Complete Siding Handbook
James E. Brumbaugh
5½ x 8¼   Hardcover   512 pp.   450 illus.
ISBN: 0-02-517880-6 $23.95

Companion to *Complete Roofing Handbook,* with step-by-step instructions and drawings on every aspect of siding: • sidewalls and siding • wall preparation • wood board siding • plywood panel and lap siding • hardboard panel and lap siding • wood shingle and shake siding • aluminum and steel siding • vinyl siding • exterior paints and stains • refinishing of siding, gutter and downspout systems • listings of professional and trade associations and siding manufacturers.

Masons and Builders Library
second edition   2 vols
Louis M. Dezettel; revised by Tom Philbin
5½ x 8¼   Hardcover   688 pp.   500 illus.
ISBN: 0-672-23401-7 $23.95

Two-volume set on practical instruction in all aspects of materials and methods of bricklaying and masonry: • brick • mortar • tools • bonding • corners, openings, and arches • chimneys and fireplaces • structural clay tile and glass block • brick walks, floors, and terraces • repair and maintenance • plasterboard and plaster • stone and rock masonry • reading blueprints.

Volume 1, Concrete, Block, Tile, Terrazzo
5½ x 8¼   Hardcover   304 pp.   190 illus.
ISBN: 0-672-23402-5 $13.95

Volume 2, Bricklaying, Plastering, Rock Masonry, Clay Tile
5½ x 8¼   Hardcover   384 pp.   310 illus.
ISBN: 0-672-23403-3 $12.95

Woodworking

Woodworking and Cabinetmaking
F. Richard Boller
5½ x 8¼   Hardcover   360 pp.   455 illus.
ISBN: 0-02-512800-0 $16.95

Compact one-volume guide to the essentials of all aspects of woodworking: • properties of softwoods, hardwoods, plywood, and composition wood • design, function, appearance, and structure • project planning • hand tools • machines • portable electric tools • construction • the home workshop • and the projects themselves – stereo cabinet, speaker cabinets, bookcase, desk, platform bed, kitchen cabinets, bathroom vanity.

Wood Furniture: Finishing, Refinishing, Repairing second edition
James E. Brumbaugh
5½ x 8¼   Hardcover   352 pp.   185 illus.
ISBN: 0-672-23409-2 $12.95

Complete, fully illustrated guide to repairing furniture and to finishing and refinishing wood surfaces for professional woodworkers and do-it-yourselfers: • tools and supplies • types of wood • veneering • inlaying • repairing, restoring, and stripping • wood preparation • staining • shellac, varnish, lacquer, paint and enamel, and oil and wax finishes • antiquing • gilding and bronzing • decorating furniture.

Maintenance and Repair

Building Maintenance second edition
Jules Oravetz
5½ x 8¼   Hardcover   384 pp.   210 illus.
ISBN: 0-672-23278-2 $9.95

Complete information on professional maintenance procedures used in office, educational, and commercial buildings: • painting and decorating • plumbing and pipe fitting

• concrete and masonry • carpentry • roofing • glazing and caulking • sheet metal • electricity • air conditioning and refrigeration • insect and rodent control • heating • maintenance management • custodial practices.

Gardening, Landscaping and Grounds Maintenance
third edition
Jules Oravetz
5½ x 8¼   Hardcover   424 pp.   340 illus.
ISBN: 0-672-23417-3 $15.95

Practical information for those who maintain lawns, gardens, and industrial, municipal, and estate grounds: • flowers, vegetables, berries, and house plants • greenhouses • lawns • hedges and vines • flowering shrubs and trees • shade, fruit and nut trees • evergreens • bird sanctuaries • fences • insect and rodent control • weed and brush control • roads, walks, and pavements • drainage • maintenance equipment • golf course planning and maintenance.

Home Maintenance and Repair: Walls, Ceilings and Floors
Gary D. Branson
8½ x 11   Paperback   80 pp.   80 illus.
ISBN: 0-672-23281-2 $6.95

Do-it-yourselfer's step-by-step guide to interior remodeling with professional results: • general maintenance • wallboard installation and repair • wallboard taping • plaster repair • texture paints • wallpaper techniques • paneling • sound control • ceiling tile • bath tile • energy conservation.

Painting and Decorating
Rex Miller and Glenn E. Baker
5½ x 8¼   Hardcover   464 pp.   325 illus.
ISBN: 0-672-23405-x $18.95

Practical guide for painters, decorators, and homeowners to the most up-to-date materials and techniques: • job planning • tools and equipment needed • finishing materials • surface preparation • applying paint and stains · decorating with coverings • repairs and maintenance • color and decorating principles.

Tree Care second edition
John M. Haller
8½ x 11 Paperback 224 pp. 305 illus.
ISBN: 0-02-062870-6 $9.95

New edition of a standard in the field, for growers, nursery owners, foresters, landscapers, and homeowners: • planting • pruning • fertilizing • bracing and cabling • wound repair • grafting • spraying • disease and insect management • coping with environmental damage • removal • structure and physiology • recreational use.

Upholstering
updated
James E. Brumbaugh
5½ x 8¼ Hardcover 400 pp. 380 illus.
ISBN: 0-672-23372-x $12.95

Essentials of upholstering for professional, apprentice, and hobbyist: • furniture styles • tools and equipment • stripping • frame construction and repairs • finishing and refinishing wood surfaces • webbing • springs • burlap, stuffing, and muslin • pattern layout • cushions • foam padding • covers • channels and tufts • padded seats and slip seats • fabrics • plastics • furniture care.

Automotive and Engines

Diesel Engine Manual fourth edition
Perry O. Black; revised by William E. Scahill
5½ x 8¼ Hardcover 512 pp. 255 illus.
ISBN: 0-672-23371-1 $15.95

Detailed guide for mechanics, students, and others to all aspects of typical two- and four-cycle engines: • operating principles • fuel oil • diesel injection pumps • basic Mercedes diesels • diesel engine cylinders • lubrication • cooling systems • horsepower • engine-room procedures • diesel engine installation • automotive diesel engine • marine diesel engine • diesel electrical power plant • diesel engine service.

Gas Engine Manual third edition
Edwin P. Anderson; revised by Charles G. Facklam
5½ x 8¼ Hardcover 424 pp. 225 illus.
ISBN: 0-8161-1707-1 $12.95

Indispensable sourcebook for those who operate, maintain, and repair gas engines of all types and sizes: • fundamentals and classifications of engines · engine parts • pistons • crankshafts • valves • lubrication, cooling, fuel, ignition, emission

control and electrical systems • engine tune-up • servicing of pistons and piston rings, cylinder blocks, connecting rods and crankshafts, valves and valve gears, carburetors, and electrical systems.

Small Gasoline Engines
Rex Miller and Mark Richard Miller
5 ½ x 8¼ Hardcover 640 pp. 525 illus.
ISBN: 0-672-23414-9 $16.95

Practical information for those who repair, maintain, and overhaul two- and four-cycle engines – with emphasis on one-cylinder motors – including lawn mowers, edgers, grass sweepers, snowblowers, emergency electrical generators, outboard motors, and other equipment up to ten horsepower: • carburetors, emission controls, and ignition systems • starting systems • hand tools • safety • power generation • engine operations • lubrication systems • power drivers • preventive maintenance • step-by-step overhauling procedures • troubleshooting • testing and inspection • cylinder block servicing.

Truck Guide Library 3 vols
James E. Brumbaugh
5½ x 8¼ Hardcover 2,144 pp. 1,715 illus.
ISBN: 0-672-23392-4 $45.95

Three-volume comprehensive and profusely illustrated reference on truck operation and maintenance.

Volume 1, Engines
5½ x 8¼ Hardcover 416 pp. 290 illus.
ISBN: 0-672-23356-8 $16.95

• Basic components • engine operating principles • troubleshooting • cylinder blocks • connecting rods, pistons, and rings • crankshafts, main bearings, and flywheels • camshafts and valve trains • engine valves.

Volume 2, Engine Auxiliary Systems
5½ x 8¼ Hardcover 704 pp. 520 illus.
ISBN: 0-672-23357-6 $16.95

• Battery and electrical systems • spark plugs • ignition systems, charging and starting systems • lubricating, cooling, and fuel systems • carburetors and governors • diesel systems • exhaust and emission-control systems.

Volume 3, Transmissions, Steering, and Brakes
5½ x 8¼ Hardcover 1,024 pp. 905 illus.
ISBN: 0-672-23406-8 $16.95

• Clutches • manual, auxiliary, and automatic transmissions • frame and suspension systems • differentials and axles, manual and power steering • front-end alignment • hydraulic, power, and air brakes • wheels and tires • trailers.

Drafting

Answers on Blueprint Reading
fourth edition
Roland E. Palmquist; revised by Thomas J. Morrisey
5½ x 8¼ Hardcover 320 pp. 275 illus.
ISBN: 0-8161-1704-7 $12.95

Complete question-and-answer instruction manual on blueprints of machines and tools, electrical systems, and architecture: • drafting scale • drafting instruments • conventional lines and representations • pictorial drawings • geometry of drafting • orthographic and working drawings • surfaces • detail drawing • sketching • map and topographical drawings • graphic symbols • architectural drawings • electrical blueprints • computer-aided design and drafting. Also included is an appendix of measurements • metric conversions • screw threads and tap drill sizes • number and letter sizes of drills with decimal equivalents • double depth of threads • tapers and angles.

Hobbies

Complete Course in Stained Glass
Pepe Mendez
8½ x 11 Paperback 80 pp. 50 illus.
ISBN: 0-672-23287-1 $8.95

Guide to the tools, materials, and techniques of the art of stained glass, with ten fully illustrated lessons: • how to cut glass • cartoon and pattern drawing • assembling and cementing • making lamps using various techniques • electrical components for completing lamps • sources of materials • glossary of terminology and techniques of stained glasswork.

VII

Macmillan Practical Arts Library
Books for and by the Craftsman

World Woods in Color
W.A. Lincoln
7 × 10   Hardcover   300 pages
300 photos
ISBN: 0-02-572350-2 $39.95

Large full-color photographs show the natural grain and features of nearly 300 woods: • commercial and botanical names • physical characteristics, mechanical properties, seasoning, working properties, durability, and uses • the height, diameter, bark, and places of distribution of each tree • indexing of botanical, trade, commercial, local, and family names • a full bibliography of publications on timber study and identification.

The Woodworker's Bible
Alf Martensson
8 × 10   Paperback   288 pages   900 illus.
ISBN: 0-02-011940-2 $12.95

For the craftsperson familiar with basic carpentry skills, a guide to creating professional-quality furniture, cabinetry, and objects d'art in the home workshop: • techniques and expert advice on fine craftsmanship whether tooled by hand or machine • joint-making • assembling to ensure fit • finishes. Author, who lives in London and runs a workshop called Woodstock, has also written. *The Book of Furnituremaking.*

Cabinetmaking: The Professional Approach
Alan Peters
8½ × 11   Hardcover   208 pages   175 illus.
(8 pp. color)
ISBN: 0-02-596200-0 $29.95

A unique guide to all aspects of professional furniture making, from an English master craftsman: • the Cotswold School and the birth of the furniture movement • setting up a professional shop • equipment • finance and business efficiency • furniture design • working to commission • batch production, training, and techniques • plans for nine projects.

The Woodturner's Art: Fundamentals and Projects
Ron Roszkiewicz
8 × 10   Hardcover   256 pages   300 illus.
ISBN: 0-02-605250-4 $24.95

A master woodturner shows how to design and create increasingly difficult projects step-by-step in this book suitable for the beginner and the more advanced student: • spindle and faceplate turning • tools • techniques • classic turnings from various historical periods • more than 30 types of projects including boxes, furniture, vases, and candlesticks • making duplicates • projects using combinations of techniques and more than one kind of wood. Author has also written *The Woodturner's Companion.*

Cabinetmaking and Millwork
John L. Feirer
7⅛ × 9½   Hardcover   992 pages
2,350 illus. (32 pp. in color)
ISBN: 0-02-537350-1 $47.50

The classic on cabinetmaking that covers in detail all of the materials, tools, machines, and processes used in building cabinets and interiors, the production of furniture, and other work of the finish carpenter and millwright: • fixed installations such as paneling, built-ins, and cabinets • movable wood products such as furniture and fixtures • which woods to use, and why and how to use them in the interiors of homes and commercial buildings • metrics and plastics in furniture construction.

Carpentry and Building Construction
John L. Feirer and Gilbert R. Hutchings
7½ × 9½   hardcover   1,120 pages
2,000 photos (8 pp. in color)
ISBN: 0-02-537360-9 $50.00

A classic by Feirer on each detail of modern construction: • the various machines, tools, and equipment from which the builder can choose • laying of a foundation • building frames for each part of a building • details of interior and exterior work • painting and finishing • reading plans • chimneys and fireplaces • ventilation • assembling prefabricated houses.